Reviews of Environmental Contamination and Toxicology

VOLUME 135

Reviews of Environmental Contamination and Toxicology

Continuation of Residue Reviews

Editor
George W. Ware

Editorial Board
F. Bro-Rasmussen, Lyngby, Denmark
D.G. Crosby, Davis, California, USA · H. Frehse, Leverkusen-Bayerwerk, Germany
H.F. Linskens, Nijmegen, The Netherlands · O. Hutzinger, Bayreuth, Germany
N.N. Melnikov, Moscow, Russia · M.L. Leng, Midland, Michigan, USA
R. Mestres, Montpellier, France · D.P. Morgan, Oakdale, Iowa, USA
P. De Pietri-Tonelli, Milano, Italy
Raymond S.H. Yang, Fort Collins, Colorado, USA

Founding Editor
Francis A. Gunther

VOLUME 135

Springer-Verlag
New York Berlin Heidelberg London Paris
Tokyo Hong Kong Barcelona Budapest

Coordinating Board of Editors

GEORGE W. WARE, *Editor*
Reviews of Environmental Contamination and Toxicology

Department of Entomology
University of Arizona
Tucson, Arizona 85721, USA
(602) 299-3735 (phone and FAX)

HERBERT N. NIGG, *Editor*
Bulletin of Environmental Contamination and Toxicology

University of Florida
700 Experimental Station Road
Lake Alfred, Florida 33850, USA
(813) 956-1151; FAX (813) 956-4631

ARTHUR BEVENUE, *Editor*
Archives of Environmental Contamination and Toxicology

4213 Gann Store Road
Hixson, Tennessee 37343, USA
(615) 877-5418

Springer-Verlag
New York: 175 Fifth Avenue, New York, NY 10010, USA
Heidelberg: 6900 Heidelberg 1, Postfach 105 280, Germany

Library of Congress Catalog Card Number 62-18595.
Printed in the United States of America.

ISSN 0179-5953

© 1994 by Springer-Verlag New York, Inc.
Copyright is not claimed for works by U.S. Government employees.
All rights reserved. This work may not be translated or copied in whole or in part without the written permission of the publisher (Springer-Verlag, 175 Fifth Avenue, New York, NY 10010, USA), except for brief excerpts in connection with reviews or scholarly analysis. Use in connection with any form of information storage and retrieval, electronic adaptation, computer software, or by similar or dissimilar methodology now known or hereafter developed is forbidden.
The use of general descriptive names, trade names, trademarks, etc. in this publication, even if the former are not especially identified, is not to be taken as a sign that such names, as understood by the Trade Marks and Merchandise Marks Act, may accordingly be used freely by anyone.

ISBN 0-387-94192-4 Springer-Verlag New York Berlin Heidelberg
ISBN 3-540-94192-4 Springer-Verlag Berlin Heidelberg New York

Foreword

International concern in scientific, industrial, and governmental communities over traces of xenobiotics in foods and in both abiotic and biotic environments has justified the present triumvirate of specialized publications in this field: comprehensive reviews, rapidly published research papers and progress reports, and archival documentations. These three international publications are integrated and scheduled to provide the coherency essential for nonduplicative and current progress in a field as dynamic and complex as environmental contamination and toxicology. This series is reserved exclusively for the diversified literature on "toxic" chemicals in our food, our feeds, our homes, recreational and working surroundings, our domestic animals, our wildlife and ourselves. Tremendous efforts worldwide have been mobilized to evaluate the nature, presence, magnitude, fate, and toxicology of the chemicals loosed upon the earth. Among the sequelae of this broad new emphasis is an undeniable need for an articulated set of authoritative publications, where one can find the latest important world literature produced by these emerging areas of science together with documentation of pertinent ancillary legislation.

Research directors and legislative or administrative advisers do not have the time to scan the escalating number of technical publications that may contain articles important to current responsibility. Rather, these individuals need the background provided by detailed reviews and the assurance that the latest information is made available to them, all with minimal literature searching. Similarly, the scientist assigned or attracted to a new problem is required to glean all literature pertinent to the task, to publish new developments or important new experimental details quickly, to inform others of findings that might alter their own efforts, and eventually to publish all his/her supporting data and conclusions for archival purposes.

In the fields of environmental contamination and toxicology, the sum of these concerns and responsibilities is decisively addressed by the uniform, encompassing, and timely publication format of the Springer-Verlag (Heidelberg and New York) triumvirate:

Reviews of Environmental Contamination and Toxicology [Vol. 1 through 97 (1962–1986) as Residue Reviews] for detailed review articles concerned with any aspects of chemical contaminants, including pesticides, in the total environment with toxicological considerations and consequences.

Bulletin of Environmental Contamination and Toxicology (Vol. 1 in 1966) for rapid publication of short reports of significant advances and discoveries in the fields of air, soil, water, and food contamination and pollution as well as methodology and other disciplines concerned with the introduction, presence, and effects of toxicants in the total environment.

Archives of Environmental Contamination and Toxicology (Vol. 1 in 1973) for important complete articles emphasizing and describing original experimental or theoretical research work pertaining to the scientific aspects of chemical contaminants in the environment.

Manuscripts for *Reviews* and the *Archives* are in identical formats and are peer reviewed by scientists in the field for adequacy and value; manuscripts for the *Bulletin* are also reviewed, but are published by photo-offset from camera-ready copy to provide the latest results with minimum delay. The individual editors of these three publications comprise the joint Coordinating Board of Editors with referral within the Board of manuscripts submitted to one publication but deemed by major emphasis or length more suitable for one of the others.

Coordinating Board of Editors

Preface

Not a day passes that any person who reads newspapers, listens to radio, or watches television is not exposed to a litany of worldwide environmental insults: acid rain resulting from atmospheric SO_2 and NOx, global warming (greenhouse effect) in relation to increased atmospheric CO_2, toxic and nuclear waste disposal, contamination of the ocean "commons," forest decline, radioactive contamination of our surroundings by nuclear power generators, and the effect of chlorofluorocarbons in reduction of the ozone layer. These represent only the most prevalent topics. In more localized disclosures, we are reminded of leaking underground fuel tanks; increasing air pollution in our cities; radon seeping into residential basements; movement of nitrates, nitrites, pesticides, and industrial solvents into groundwater supplies; and contamination of our food and animal feeds with pesticides, industrial chemicals, and bacterial toxins. It then comes as no surprise that ours is the first generation of mankind to have become afflicted with the pervasive and acute (but perhaps curable) disease appropriately named "chemophobia," or fear of chemicals.

There is abundant evidence, however, that most chemicals are degraded or dissipated in our not-so-fragile environment, despite efforts by environmental ethicists and the media to convince us otherwise. But for most scientists involved in reduction of environmental contaminants, there is indeed room for improvement in virtually all spheres.

For those who make the decisions about how our planet is managed, there is an ongoing need for continual surveillance and intelligent controls, to avoid endangering the environment, wildlife, and the public health. Ensuring safety-in-use of the many chemicals involved in our highly industrialized culture is a dynamic challenge, for the old established materials are continually being displaced by newly developed molecules more acceptable to environmentalists, federal and state regulatory agencies, and public health officials.

Environmentalism has become a worldwide political force, resulting in multi-national consortia emerging to control pollution and in the maturation of the environmental ethic. Will the new politics of the next century be a consortium of technologists and environmentalists or a confrontation? These matters are of genuine concern to governmental agencies and legislative bodies around the world, for many chemical incidents have resulted from accidents and improper use.

Adequate safety-in-use evaluations of all chemicals persistent in our air, foodstuffs, and drinking water are not simple matters, and they incorporate the judgments of many individuals highly trained in a variety of complex biological, chemical, food technological, medical, pharmacological, and toxicological disciplines.

We intend that *Reviews of Environmental Contamination and Toxicology* will continue to serve as an integrating factor both in focusing attention on those matters requiring further study and in collating for variously trained readers current knowledge in specific important areas involved with chemical contaminants in the total environment. Previous volumes of *Reviews* illustrate these objectives.

Because manuscripts are published in the order in which they are received in final form, it may seem that some important aspects of analytical chemistry, bioaccumulation, biochemistry, human and animal medicine, legislation, pharmacology, physiology, regulation, and toxicology have been neglected at times. However, these apparent omissions are recognized, and pertinent manuscripts are in preparation. The field is so very large and the interests in it are so varied that the Editor and the Editorial Board earnestly solicit authors and suggestions of underrepresented topics to make this international book series yet more useful and worthwhile.

Reviews of Environmental Contamination and Toxicology attempts to provide concise, critical reviews of timely advances, philosophy, and significant areas of accomplished or needed endeavor in the total field of xenobiotics in any segment of the environment, as well as toxicological implications. These reviews can be either general or specific, but properly they may lie in the domains of analytical chemistry and its methodology, biochemistry, human and animal medicine, legislation, pharmacology, physiology, regulation, and toxicology. Certain affairs in food technology concerned specifically with pesticide and other food-additive problems are also appropriate subjects.

Justification for the preparation of any review for this book series is that it deals with some aspect of the many real problems arising from the presence of any foreign chemical in our surroundings. Thus, manuscripts may encompass case studies from any country. Added plant or animal pest-control chemicals or their metabolites that may persist into food and animal feeds are within this scope. Food additives (substances deliberately added to foods for flavor, odor, appearance, and preservation, as well as those inadvertently added during manufacture, packing, distribution, and storage) are also considered suitable review material. Additionally, chemical contamination in any manner of air, water, soil, or plant or animal life is within these objectives and their purview.

Normally, manuscripts are contributed by invitation, but suggested topics are welcome. Preliminary communication with the Editor is recommended before volunteered review manuscripts are submitted.

Department of Entomology G.W.W.
University of Arizona
Tucson, Arizona

Table of Contents

Foreword v
Preface vii

Prediction and Monitoring of the Carcinogenicity of Polycyclic
Aromatic Compounds (PACs) 1
    GLEN R. SHAW and DES W. CONNELL

Interactions of Pesticides and Metal Ions with Soils:
Unifying Concepts 63
    DONALD S. GAMBLE, COOPER H. LANGFORD, and
    G.R. BARRIE WEBSTER

Turtles as Monitors of Chemical Contaminants
in the Environment 93
    LINDA MEYERS-SCHÖNE and BARBARA T. WALTON

Index 155

Prediction and Monitoring of the Carcinogenicity of Polycyclic Aromatic Compounds (PACs)

Glen R. Shaw* and Des W. Connell†

Contents

I. Introduction	1
II. Mechanisms of Chemical Carcinogenesis with PACs	5
A. Principles of Chemical Carcinogenesis	5
B. Metabolic Activation of PACs	7
C. Formation of Adducts by PACs	13
D. Relationship Between Adduct Formation and Carcinogenicity	15
III. Biomonitoring for Carcinogenicity	20
A. Use of DNA Adducts Formed in Human Tissues	20
B. Use of Protein Adducts Formed in Human Tissues	25
C. Comparison of the Application of DNA Adduct and Protein Adduct Monitoring	27
IV. Quantitative Structure Activity Relationships (QSARs)	27
A. Mechanistic Approach	27
B. Nonmechanistic Approach	39
C. Combined Approach for QSARs to Predict Carcinogenicity of PACs	44
V. Conclusions	45
Summary	46
References	48

I. Introduction

The manufacture and use of synthetic chemicals and the production and utilization of energy are primarily responsible for an increase in the worldwide production and distribution of chemicals. Health authorities are concerned about the extent to which response to low levels of carcinogens plays a role in the etiology of cancer (Rall 1990). The International Agency for Research on Cancer (IARC) Monographs Supplement 7 (IARC 1987) lists 39 chemicals or groups of chemicals and 11 industrial processes known to cause cancer in humans. Of these, five processes, six substances, and three chemicals listed in Table 1 involve polycyclic aromatic compounds (PACs).

The history of association of cancer with chemicals began when Percival Pott (1775) recognized the connection between scrotal cancer in English

---

*National Research Centre for Environmental Toxicology, Coopers Plains, Queensland 4108, Australia.
†Government Chemical Laboratory, Coopers Plains, Queensland 4108, Australia.

© 1994 by Springer-Verlag New York, Inc.
Reviews of Environmental Contamination and Toxicology, Vol. 135.

Table 1. Chemical Substances and Processes Classified
as Carcinogenic (IARC Group 1)[a]

Chemical Substances	Chemical Processes
coal tar pitches	aluminum production
coal tar	coal gasification
mineral oils (untreated or mildly treated)	coke production
	rubber industry
shale oils	
soots	
tobacco smoke	
4-aminobiphenyl	
benzidine	
2-naphthylamine	

[a]Only substances and processes that involve PACs are listed.

chimney sweeps and exposure to soot. Pott's proposal for a causal relationship between cancer and soot was supported when Butlin (1892) reported the relative lack of scrotal cancer in chimney sweeps in Europe, who had less exposure compared with those in England.

Experimental studies of chemical carcinogenesis first began in 1915 when Yamagawa and Ichikawa induced skin carcinomas on the ears of rabbits with repeated application of coal tar over extended periods (Pitot 1990). The first pure chemical carcinogen was recognized in 1930 when Kennaway and Hiegar (1930) synthesized dibenz[a,h]anthracene and determined its carcinogenicity. It was found that the fluorescence spectrum of dibenz[a,h]anthracene did not correspond exactly to that of the carcinogenic components of tars. An investigation of other carcinogenic components of coal tars by Cook et al. (1933) led to the discovery of benzo[a]pyrene as a major carcinogenic component of coal tars. Since these pioneering studies, much research has been undertaken on the carcinogenic properties of PACs. In excess of 30 parent PACs and several hundred alkyl derivatives of PACs have been reported to exhibit some carcinogenic effects (Bjorseth and Becher 1986; Hemminki et al. 1990). In addition, it has been shown that at least 75% of the total carcinogenic effect of condensates from exhausts of gasoline and diesel engines, coal combustion, and cigarette smoke can be attributed to PACs containing four or more rings (Grimmer et al. 1991).

Table 2 presents IARC classifications of the carcinogenic activity and degrees of evidence for carcinogenicity to humans and experimental animals for selected PACs. While there is unequivocal evidence for the carcinogenicity of many PACs in animals, a number of PACs are classified as probable human carcinogens due to the fact that human exposure to this class of chemical is always in combination with many other chemicals (Hemminki et al. 1990; IARC 1984a, 1984b, 1985). The epidemiological

Table 2. Carcinogenic Activity Evaluation of Selected PACs[a]

Compound	Evidence for Carcinogenicity[b]	Overall Evaluation[c]
5-aminoacenaphthene	I	3
2-aminoanthraquinone	L	3
1-amino-3-methylanthraquinone	L	3
antanthrene	L	3
anthracene	I	3
benz[a]acridine	I	3
benz[c]acridine	L	3
benz[a]anthracene	S	2A
benzo[b]fluoranthene	S	2B
benzo[j]fluoranthene	S	2B
benzo[k]fluoranthene	S	2B
benzo[b]fluorene	I	3
benzo[c]fluorene	I	3
benzo[g,h,i]perylene	I	3
benzo[c]phenanthrene	I	3
benzo[a]pyrene	S	2A
benzo[e]pyrene	I	3
chrysene	L	3
cyclopenta[c,d]pyrene	L	3
dibenz[a,h]acridine	S	2B
dibenz[a,j]acridine	S	2B
dibenz[a,c]anthracene	L	3
dibenz[a,h]anthracene	S	2A
dibenz[a,j]anthracene	L	3
7-H-dibenzo[c,g]carbazole	S	2B
dibenzo[a,e]fluoranthene	L	3
dibenzo[h,r,s,t]pentaphene	L	3
dibenzo[a,e]pyrene	S	2B
dibenzo[a,h]pyrene	S	2B
dibenzo[a,i]pyrene	S	2B
dibenzo[a,l]pyrene	S	2B
1,4-dimethylphenanthrene	I	3
1,8-dinitropyrene	I	3
fluoranthene	I	3
fluorene	I	3
indeno[1,2,3-c,d]pyrene	S	2B
1-methylchrysene	I	3
2-methylchrysene	L	3
3-methylchrysene	L	3
4-methylchrysene	L	3
5-methylchrysene	S	2B
6-methylchrysene	L	3
2-methylfluoranthene	L	3

(continued)

Table 2. (*Continued*)

Compound	Evidence for Carcinogenicity[b]	Overall Evaluation[c]
3-methylfluoranthene	I	3
1-methylphenanthrene	I	3
6-nitrobenzo[a]pyrene	I	3
6-nitrochrysene	I	3
3-nitrofluoranthene	I	3
1-nitropyrene	L	3
perylene	I	3
phenanthrene	I	3
pyrene	I	3

[a]According to IARC (1987).
[b]For humans: no adequate data are available. For animals: I, inadequate evidence; L, limited evidence; S, sufficient evidence.
[c]1, Carcinogenic to humans, but no PACs are classified in this category. 2A, Probably carcinogenic to humans. 2B, Possibly carcinogenic to humans. 3, Not classifiable as to its carcinogenicity to humans. 4, Probaly not carcinogenic to humans.

evidence of cancer risk from occupations such as chimney sweeps, aluminum workers, coke oven and coal gasification workers, iron and steel foundary workers, and asphalt and tar workers, however, shows that carcinogenicity is very likely to be caused, in part, by PACs (Bjorseth and Becker 1986; Hemminki et al. 1990; IARC 1984a, 1984b, 1985).

The PACs as a class of compounds consist of two or more fused aromatic rings in various structural configurations. The polycyclic aromatic hydrocarbons (PAHs) are the best represented members of the group, which also includes S, O, and N heterocyclics (Fig. 1). The PACs may contain substituent groups such as alkyl, hydroxyl, amino, nitro, and others. The PACs can be divided into kata-annelated or pericondensed systems. The kata-annelated systems contain certain carbon atoms that are the centers of two rings such as anthracene, whereas the pericondensed systems contain some carbon atoms that are the centers of three linked rings such as pyrene (Fig. 1).

Most PACs are solids at room temperature and possess relatively low vapor pressures and water solubilities. The kata-annelated PACs are generally less water-soluble than corresponding pericondensed compounds. The physical and chemical properties of PACs are controlled to a significant extent by the conjugated π electron system of these compounds. The presence of high-energy π-bonding orbitals and low-energy π^*-antibonding orbitals in PACs causes the absorption of visible or UV radiation by the transition of an electron from the π to π^* orbital. This results in the characteristic colors, UV absorption, and fluorescence of PACs.

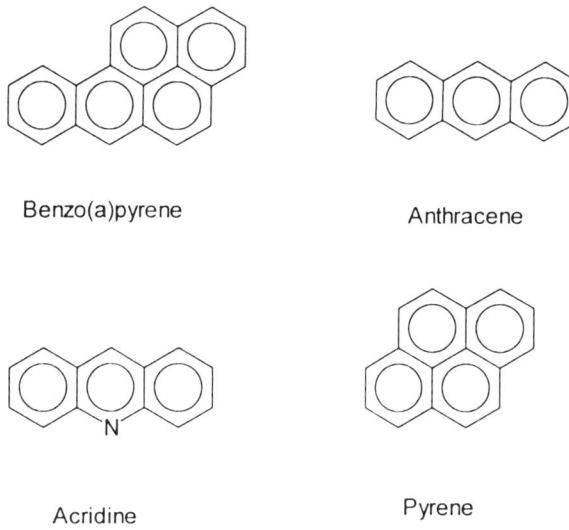

Fig. 1. Structures of some typical PACs.

The objective of this review is to evaluate chemical carcinogenesis in relation to PACs, including the mechanisms of metabolite formation and metabolic activation of carcinogenic PACs. The covalent binding of the active metabolites to DNA and proteins with the subsequent formation of adducts is also reviewed and the relationships between the adducts and the formation of cancer considered. Finally, the potential for the development of quantitative structure activity relationships (QSARS) to predict adduct formation and carcinogenicity of PACs is investigated.

II. Mechanisms of Chemical Carcinogenesis with PACs
A. Principles of Chemical Carcinogenesis

A mutagen is a substance that produces a mutation or alteration in the genetic material, whereas a carcinogen produces neoplasia (a tumor or cancer). It is generally accepted that nearly all carcinogens are mutagens whereas not all mutagens are carcinogens. It is now well known that many classes of PACs exhibit mutagenic activity (Braun et al. 1987; Haugen et al. 1981; Maher et al. 1987; Royer et al. 1983; Wild 1990). Many of the higher-molecular-weight PAHs exhibit mutagenicity in the presence of postmitochondrial supernatent (PMS) prepared from Aroclor-1254-induced rat liver. On the other hand, the PACs with carbonyl, ethenic, nitro, or amino functionalities and the nitrogen-containing heterocyclic PACs generally exhibit mutagenic activity in the absence of PMS (Austin et al. 1985; Wornat et al. 1990).

Mechanistic theories of chemical carcinogenesis were initiated by Pullman and Pullman (1955), who considered the relationships between electronic structure and carcinogenic activity of PAHs. Since then, many mechanistic theories have been developed. In the case of PACs, it is accepted that covalent binding of metabolically activated forms to DNA is the first stage in chemical carcinogenesis (Pitot 1990; Wilson et al. 1988). The DNA bears the genetic information required for expressing all cellular characteristics, including regulation of growth and cellular function. These latter chatacteristics are examples of processes that are disrupted by carcinogenesis.

With most PACs, metabolic activation must occur before covalent binding can take place. Many PACs can be considered "procarcinogens" that require metabolism to "proximate" and finally "ultimate" carcinogens. The cytochrome P-450-dependent enzyme system present in the endoplasmic reticulum of mammalian cells and most abundant in the liver catalyzes the addition of oxygen to many PACs. This converts these compounds of low water solubility into more polar and water-soluble compounds as part of the mammalian detoxification system. Although most of these oxygenated products are excreted, some metabolites have such high reactivity that adducts are formed with nucleic acids and proteins.

It has now been recognized that chemical carcinogenesis is a multistage process and the Moolgavkar-Venzon-Knudson (MVK) two-stage model can be used to quantitatively describe carcinogenesis (Walker 1989). However, this two-stage model has been modified to comprise three stages, these being initiation, promotion, and progression (Barrett 1992; Kinzel et al. 1986).

The initiation process involves the binding of chemicals to DNA, causing alterations in the genome. This is irreversible, but the effectiveness of initiation depends on the relationship between adduct formation and cellular replicative DNA synthesis and cell division (Pitot 1990; Ying et al. 1982).

The second stage of carcinogenesis is tumor promotion. Promoters are chemicals that enhance the effects of carcinogens when administered subsequently to the initiators. This stage in the progression of carcinogenicity is usually reversible (Glauert et al. 1986; Pitot 1990; Takematsu et al. 1983). In addition, it has been found that in hepatocarcinogenesis and epidermal carcinogenesis, focal lesions that disappear on removal of the promoting agent can reappear on administration (Hendrick et al. 1986; Reddy et al. 1987). The relative potency of promoters has been related to their effectiveness in increasing the progeny of initiated cell populations (Pitot 1990). Some chemicals possess promoting as well as initiating ability and can be more strongly carcinogenic than expected. For instance, it has been noted (Neumann et al. 1990) that 2-acetylaminofluorene (AAF) is capable of producing liver tumors without an additional promoter.

The third recognized phase of multistage carcinogenesis is progression. It is in this stage that irreversible benign and/or malignant neoplasms are observed in affected organisms (Barrett 1992; Pitot 1990; Schulte-Hermann 1985). Progression has been defined as "that stage of carcinogenesis exhibiting measurable (by recombinant DNA technology or related methods) and/ or morphologically identifiable (karyotypic) changes in the structure of the cell genome" (Pitot 1986). These changes are related to increased growth rate, invasiveness, metastatic capability, and biochemical changes in the neoplastic cell (Pitot 1990). It has been shown that the more malignant cancers are associated with larger numbers of genetic alterations and that these alterations are considered to be a direct reflection of the multiple steps of chemical carcinogenesis (Sugimura et al. 1992).

The mechanistic role of oncogenes and tumor suppressor genes is important in multistage carcinogenesis. Chemical carcinogens cause alterations in cellular oncogenes that are known to be targets for carcinogens. Over 50 such oncogenes have currently been identified (Walker 1989). The oncogenes in their normal form are termed "proto-oncogenes" whose functions in cell regulation include regulation of gene expression by direct binding to DNA. The cellular proto-oncogenes are present in the DNA of all cells and are essential for regulation of cell growth (Bos and van Kreyl 1992; Risse et al. 1990; Walker 1989). These proto-oncogenes are activated by various mechanisms, including point mutations that convert them to cancer-causing oncogenes (Klein 1988). Amplification of proto-oncogenes has been shown to occur in the development of many known tumors (Anderson et al. 1992).

Tumor suppression genes, in contrast to oncogenes, must become inactivated, resulting in loss of function, to allow neoplastic transformation to occur. Significant information has become available on the role of cellular oncogenes in chemical carcinogenesis, and recently more information has become available on the interaction of carcinogenic chemicals with tumor suppressor genes (Barrett 1992; Harris 1992; Walker 1989). Predisposition to certain cancers appears to result from the inheritance of a single mutated tumor suppression gene (Evans and Prosser 1992). In addition to oncogenes and tumor suppression genes, genes that accelerate cell proliferation and metastasis are also associated with multistage chemical carcinogenesis (Sugimura et al. 1992).

DNA repair also plays a significant role in the chemical carcinogenesis process (Stewart 1992). For instance, before replication, error-free repair can occur, thus effectively removing the carcinogen adduct from the DNA, or error-prone repair can occur. After replication, repair can still occur successfully or mispairing of adducted bases can occur, leading to mutation. Preferential repair of certain DNA lesions can lead to persistent lesions that therefore have a bias toward mutagenesis (Scicchitano and Hanawalt 1992).

B. Metabolic Activation of PACs

1. Polycyclic Aromatic Hydrocarbons (PAHs). It is well known that many PACs must be metabolically activated to electrophilic intermediates in order to produce carcinogenicity (Hall and Grover 1990; Haugen et al.

1988). As a detoxification mechanism, enzyme systems convert PAHs and other PACs to hydroxylated primary metabolites, are then conjugated to form secondary metabolites, and are subsequently excreted from the system. The conjugates can be glucuronides and sulfates, as well as a variety of other substances. The process of formation of primary metabolites results in the production of highly reactive intermediates (Hall and Grover 1990).

The enzyme system primarily responsible for PAH metabolism is the mixed-function oxidase system, which requires NADH or NADPH and molecular oxygen to convert the nonpolar PAHs into the polar hydroxy and epoxy derivatives (Dipple et al. 1984; Hall and Grover 1990; Mallet et al. 1991). The terminal oxidase is cytochrome P-450. Cytochrome P-450 is composed of a group of inducible microsomal mono-oxygenases with various substrate specificities. Cytochrome P-4501A1 and 1A2 plus IIIA3 and IIIA4 are the forms induced by PAHs (Friedberg et al. 1990; Scharping et al. 1992).

It has been determined that epoxides are the major intermediates in the oxidative metabolism of aromatic double bonds (Hall and Grover 1990). The epoxides, however, are quite reactive and enzymatically metabolized to other compounds such as diols and phenols that can then be conjugated with glutathione or sulfate. The formation of diols is enzymatically catalyzed by microsomal epoxide hydrolases that are induced, but not to the same extent as the mono-oxygenases (Glatt et al. 1984; Hall and Grover 1990; Thomas et al. 1990). These diols possess asymmetric centers and so exist in enantiomeric forms that can exhibit different reactivities. Pathways involved in the metabolism of benzo[*a*]pyrene (BaP) to diols and other hydroxylated derivatives are shown in Fig. 2.

The conjugation of PAH epoxides with glutathione is mediated by cytosolic enzymes, the glutathione transferases that consist of a family of dimeric isoenzymes whose individual members are composed of various combinations of two monomers (Hall and Grover 1990; Jakoby et al. 1984; Mannervik and Jensson 1982). The glutathione transferases can catalyze conjugation with simple epoxides plus diol- and triol-epoxides derived from PAHs, although different specificities exist between different epoxides (Hall and Grover, 1990; Hodgson et al. 1986; Jernstrom et al. 1985). The conjugation of PAH epoxides with glutathione is regarded as a true detoxification reaction (Hall and Grover 1990).

The epoxides that are not conjugated with glutathione are converted into phenols and diols as mentioned above. These PAH metabolites, however, are sometimes not sufficiently polar to be excreted and are therefore conjugated with glucuronic or sulfuric acids to enable excretion to occur (Hall and Grover 1990). In addition to conjugation, the hydroxylated derivatives of PAHs may undergo a number of oxidation and hydroxylation reactions. These include the conversion of phenols to phenol-epoxides and subsequently to diphenols and triols, diols to tetrols and diol-epoxides, and triols to

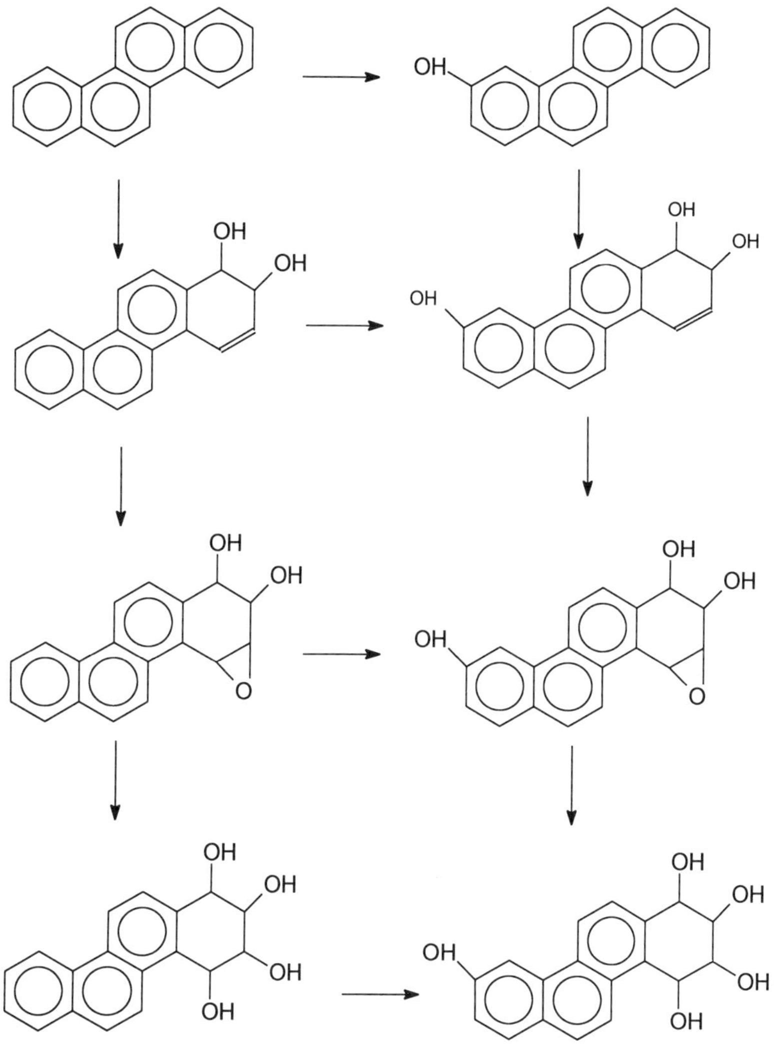

Fig. 2. Pathways involved in the metabolism of chrysene (Hall and Grover 1990).

triol-epoxides and pentols. For example, the metabolism of chrysene through to polyhydroxy compounds is shown in Fig. 2.

Metabolic activation of PAHs proceeds via the cytochrome P-450 enzymatic system as previously described. One of the first mechanisms proposed for the formation of activated electrophilic intermediates involved the formation of simple K region epoxides (Sims and Grover 1974). The K region illustrated in Fig. 3 for benz[a]anthracene is distinguished by a relatively high π electron density, producing high reactivity in this region. It was

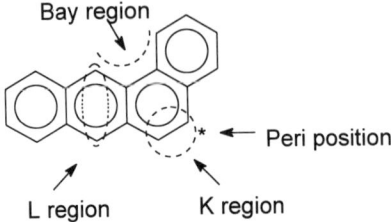

Fig. 3. Regions of activity in benz[a]anthracene.

later recognized, however, that nucleic acid adducts formed with K region epoxides were not identified in those formed in tissues treated with the parent PAHs (Baird et al. 1973, 1975). Further research led to the identification of BaP-7,8-diol-9,10-oxides as the activated forms of BaP that bound covalently to DNA (Daudel et al. 1975; Sims and Grover 1974). It is now recognized that PAHs are activated by the formation of vicinal diol-epoxides and that in most cases the diol-epoxides are formed adjacent to a bay region. A bay region is simply the region produced by the angular addition of a benzene ring to a linear portion of a PAH, as illustrated with benz[a]anthracene shown in Fig. 3. Figure 4 shows the difference between bay-region dihydrodiol-epoxides and nonbay-region dihydrodiol-epoxides of BaP.

The other main region of activity in PAH molecules is the "L" region (see Fig. 3). The L region of a PAH molecule features localization of π electrons across *para* positions, e.g., the 7,12 positions of benz[a]anthracene. The presence of an L region in a PAH was recognized by Pullman and Pullman (1955) as being responsible for the absence of carcinogenic activity in certain PAHs. The high reactivity of L regions caused by high electron density can be reduced by substitution, such as that in 7,12-dimethylbenz[a]anthracene, which is a potent carcinogen as compared to the unsubstituted benz[a]anthracene (Dipple et al. 1984).

In addition, it has been found that methyl substitution at the bay-region carbon atom adjacent to the angular ring but not on the angular ring enhances carcinogenicity, whereas substitution at the "Peri" position, i.e., the carbon atom that is opposite the bay region and adjacent to the angular ring (see Fig. 4), reduces carcinogenic activity (Seybold 1986). It can be seen in Fig. 5 that high carcinogenic activity occurs with 12-methylbenz[a]anthracene, 5-methylchrysene, and 11-methylbenzo[a]pyrene, whereas 6-methylchrysene, 5-methylbenz[a]anthracene, and 6-methylbenzo[a]pyrene are either inactive or weakly carcinogenic.

The stereochemistry of activated intermediates is important with respect to the carcinogenicity of PAHs, as a pair of enantiomers may exhibit large differences in activity and mechanisms of metabolism. The metabolism of PAHs is stereoselective due to stereoselective epoxidation of the PAH facilitated by cytochrome P-450, followed by regioselective hydration catalyzed by microsomal epoxide hydrolase (Mallet et al. 1991; Platt et al.

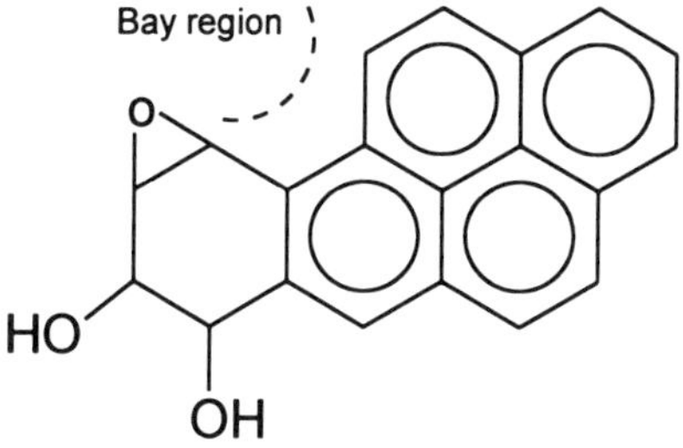

A. Bay region dihydrodiol epoxide

(7,8-Dihydroxy-9,10-epoxy-7,8,9,10-tetrahydrobenzo[a]pyrene)

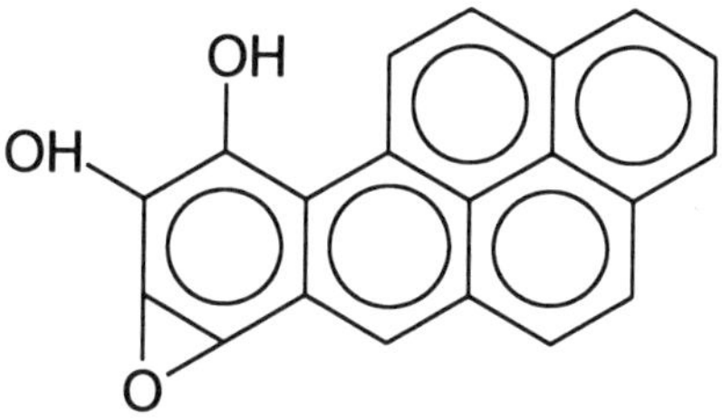

B. Nonbay-region dihydrodiol epoxide

(9,10-Dihydroxy-7,8-epoxy-7,8,9,10-tetrahydrobenzo[a]pyrene)

Fig. 4. Structures of (a) bay-region and (b) nonbay-region dihydrodiol-epoxides of benzo[a]pyrene.

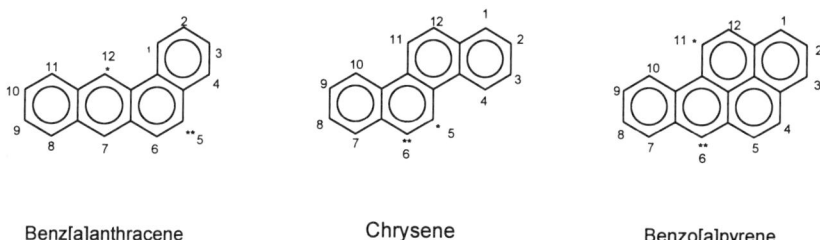

Benz[a]anthracene   Chrysene   Benzo[a]pyrene

Fig. 5. Effects of methyl substitution on the carcinogenicity of PAHs (*Producing carcinogenic activity; **not producing carcinogenic activity).

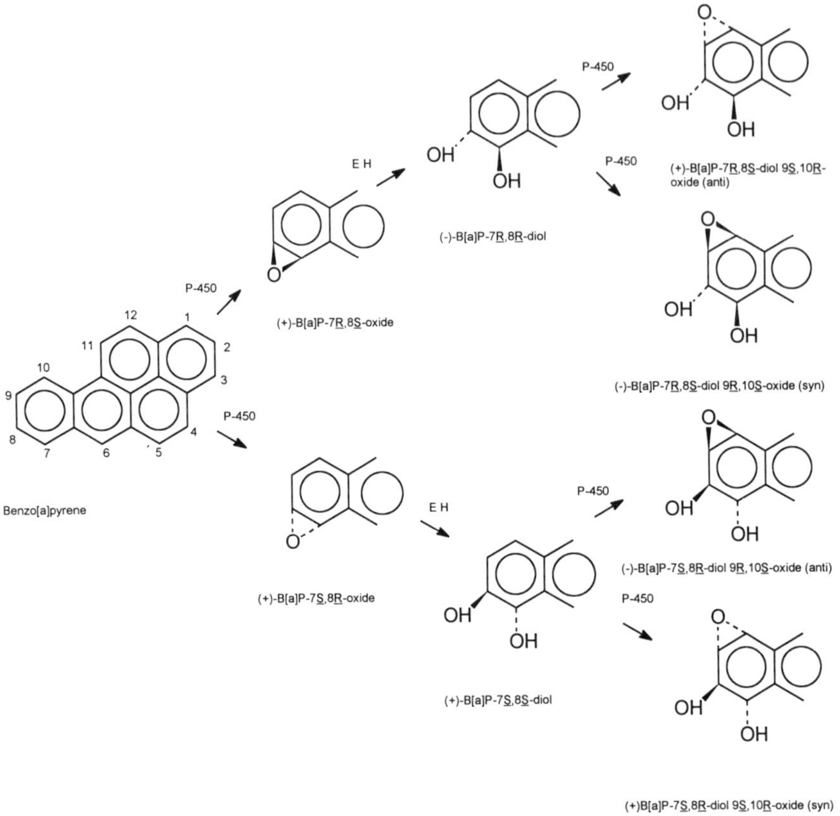

Fig. 6. Stereoselective metabolism of benzo[a]pyrene.

1990). Diol-epoxides can exist in both *syn-* or *anti-*forms, and furthermore, two enantiomers of both the *syn-* and *anti-*diol-epoxides exist, producing four stereoisomers for each diol-epoxide. It has been found that the *anti-*stereoisomers are preferentially formed compared with the *syn-*isomers (Hall and Grover 1990; Platt et al. 1990; Weyand et al. 1993). Figure 6 shows the metabolic activation of benzo[a]pyrene and the four stereoisomers formed. A comprehensive discussion on the nomenclature of PAH metabolites has been presented by Dipple et al. (1984).

2. Nitrogen Containing PACs (NPACs). The NPACs considered in this review are the nitroaromatic hydrocarbons and aromatic amines. These two classes of PACs must also be metabolized into reactive electrophiles to produce carcinogenic activity, as described in the previous section for the PAHs.

The metabolic activation of aromatic amines usually involves an initial

N-oxidation to N-hydroxyarylamines (Beland and Kadlubar 1990; Debnath et al. 1991; King et al. 1990; Roy et al. 1991b). This initial oxidation is mediated by cytochrome P-450 isozymes. The initial activation of nitroaromatic hydrocarbons also involves the formation of an N-hydroxyarylamine and is catalyzed by cytochrome P-450. The nitroaromatics and aromatic amines with three or more aromatic rings may also be metabolized to dihydrodiol-epoxides by the same activation mechanisms as PAHs. Figure 7 shows metabolic activation pathways for 6-aminochrysene and 6-nitrochrysene.

C. Formation of Adducts by PACs

1. Polycyclic Aromatic Hydrocarbons. The metabolic activation of bay-region PAHs to dihydrodiol-epoxides is the first step in the formation of DNA adducts. These electrophilic intermediates covalently bind to nucleophilic centers in biological molecules such as DNA and hemoglobin. Binding can occur at multiple nucleophilic centers on DNA, but adduct formation by activated PAH metabolites predominately occurs at the N-2 position of guanine (Chadha et al. 1989; ECETOC 1989) (Fig. 8).

The binding of dihydrodiol-epoxides to DNA bases involves epoxide ring opening and subsequent formation of a carbonium ion intermediate that then covalently binds to nucleophilic sites on the DNA bases (Stowers and Anderson 1985) (Fig. 8). It should be noted that this adduct is in an *anti* configuration and that the *anti* configuration adducts are the predominant adducts of PAHs observed (Hall and Grover 1990; Rojas and Alexandrov 1986) (see structure "a" in Fig. 6). Research has indicated that physical binding of PAH diol-epoxides to DNA can in some cases enhance adduct formation by these compounds (Paulius et al. 1986; Prakash et al. 1988).

2. Aromatic Amines and Nitroaromatics. The formation of DNA adducts by the metabolically activated intermediates of arylamines and nitroaromatics involves the covalent binding of either hydroxylamines or diol-epoxides to bases in DNA. Hydroxylamines may be converted to nitrenium ions as reactive intermediates that then react readily with nucleophilic centers on DNA, in particular C-8, N-2, and O-6 of guanine and C-8 and N-6 of adenine (Fig. 9). Sulfation or acetylation of the hydroxylamine also occurs (Beland and Kadlubar 1990; Patton et al. 1986; Roy et al. 1991a, 1991b; Silverman and Lowe 1986).

The metabolic activation of certain nitro or amino PAHs can cause the formation of adducts of both the N-hydroxylamine and diol-epoxide type (Beland and Kadlubar 1990; Roy et al. 1991b). Figure 9 shows the metabolic activation and formation of various adducts with 6-nitrochrysene and 6-aminochrysene. Preferential formation of the various adducts depends on the specific metabolic pathways for activation, which in turn has been

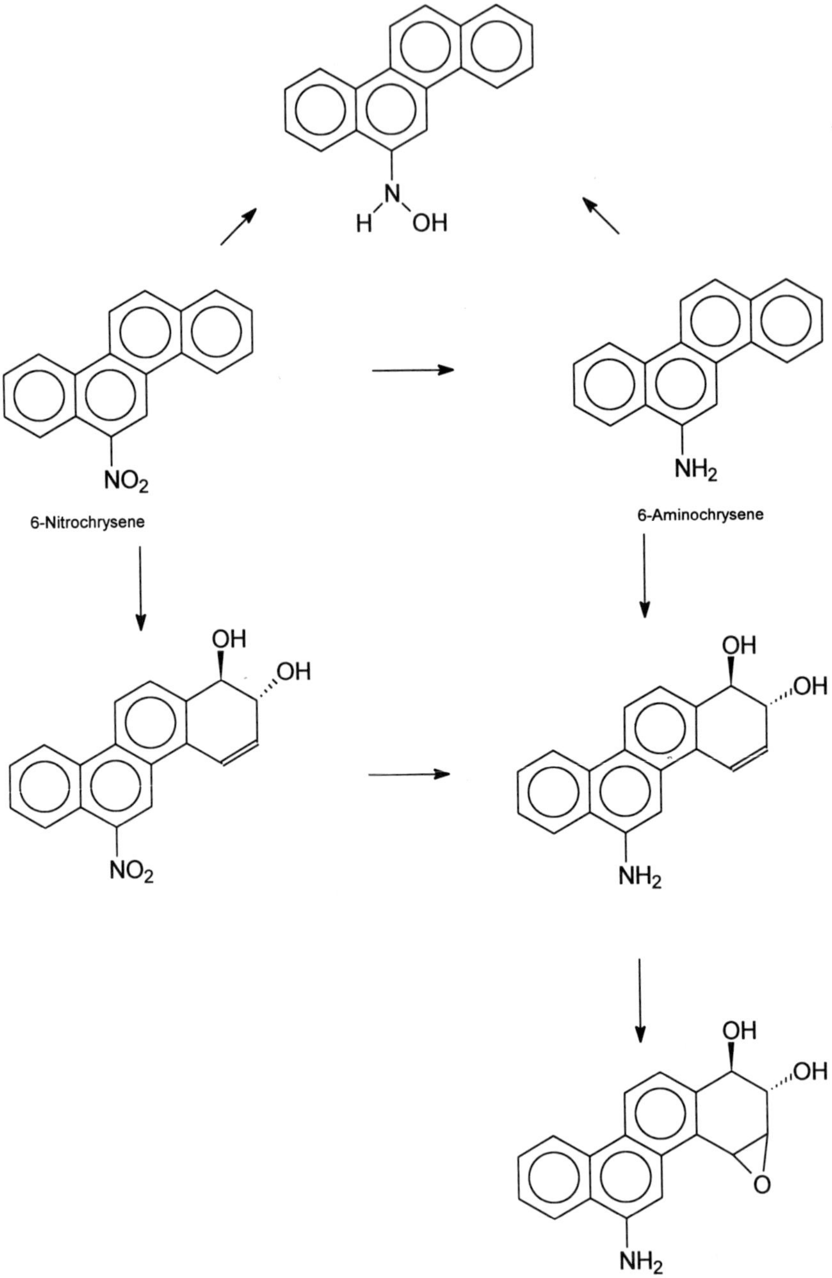

Fig. 7. Metabolic activation pathways of 6-aminochrysene and 6-nitrochrysene (Beland and Kadlubar 1990).

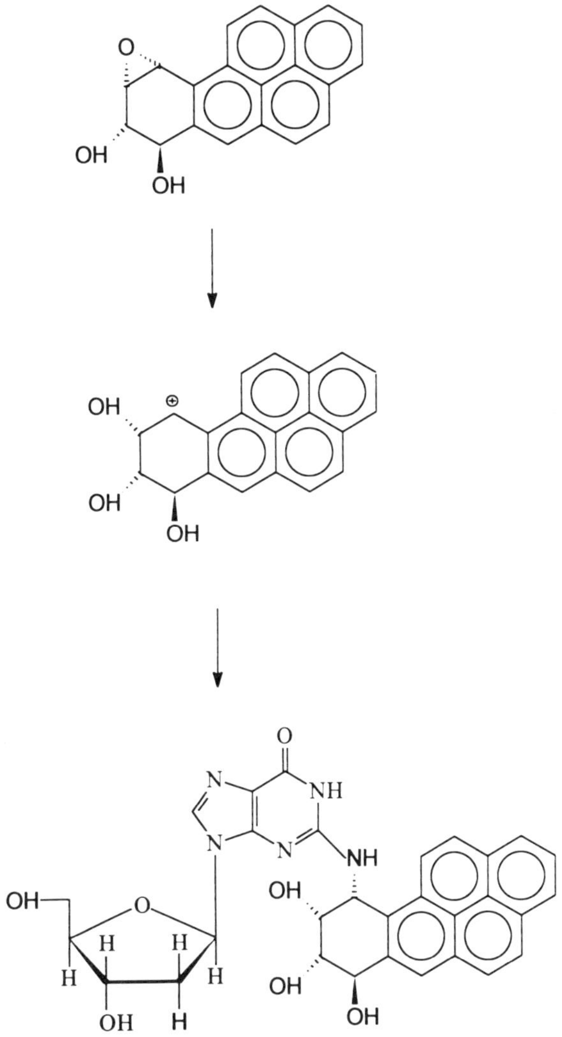

Fig. 8. DNA adduct formation between (+)-*anti*-benzo[*a*]pyrene diol-epoxide and the N2 position of deoxyguanosine.

found to depend on the particular cytochromes P-450 that are present (Beland and Kadlubar 1990; Delcos et al. 1988; Orzechowski et al. 1992).

D. Relationship Between Adduct Formation and Carcinogenicity

The relationship between adduct formation and carcinogenesis is complex and, in addition, only certain adducts formed by PACs cause tumor formation (ECETOC 1989; Swenberg et al. 1985). It has been postulated (Geactinov 1988) that highly tumorigenic diol-epoxides such as (+)*anti*-benzo[*a*]pyrene-diol-epoxide, 3-methylcholanthrene-9,10-diol-7,8-oxide, and

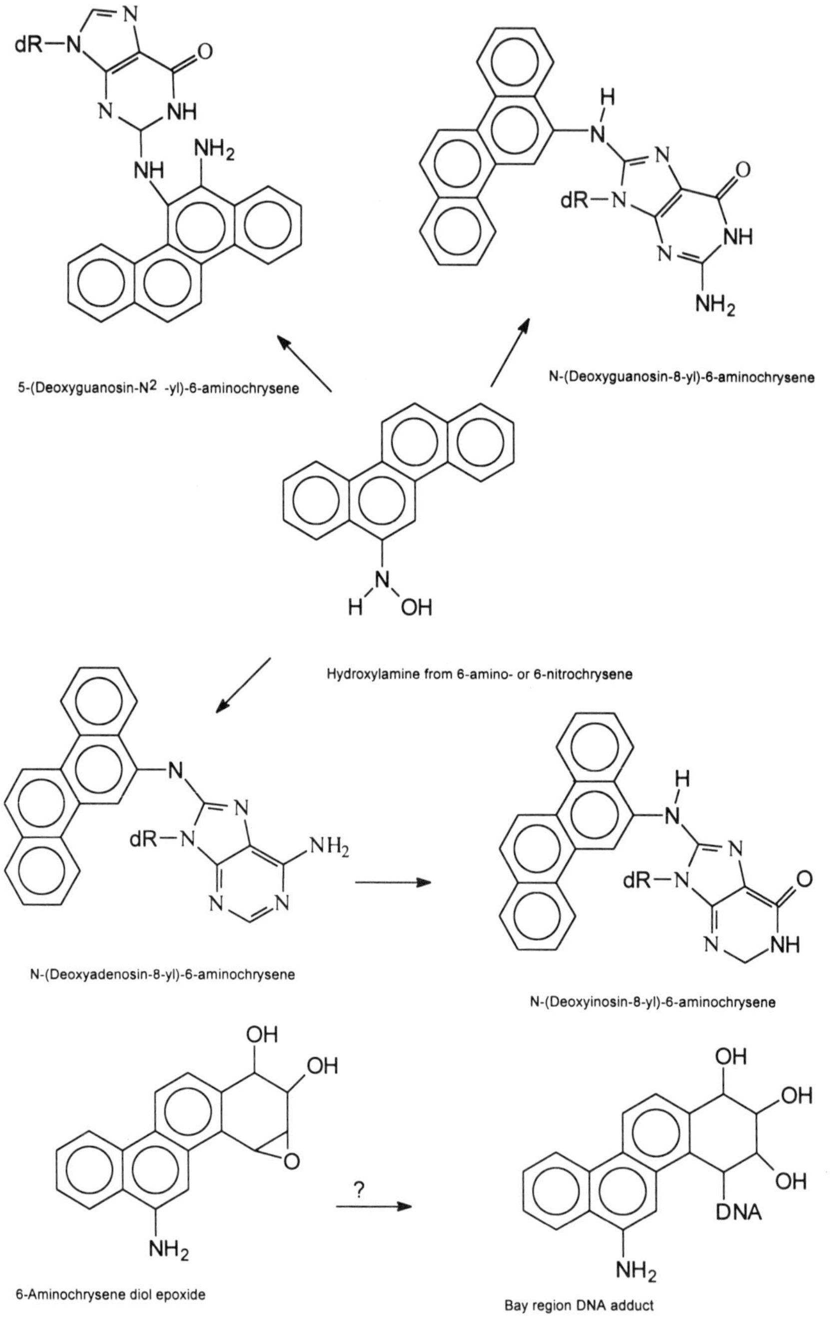

Fig. 9. Formation of DNA adducts with activated metabolites of 6-nitrochrysene and 6-aminochrysene (Beland and Kadlubar 1990).

5-methylchrysene-1,2-diol-3,4-oxide undergo a reorientation upon covalent binding to sites on DNA such as the exocyclic amino group of guanine. The conformation of these adducts is consistent with having external orientation in comparison with other adducts (Fig. 10) that have stacking between the aromatic portions of the PACs and bases, and little change in the conformations of the PAC molecules.

Multistage chemical carcinogenesis is represented diagrammatically in Fig. 11. In this model, the initiation phase includes exposure to the PAC, transport to the target cell, activation to the ultimate carcinogenic metabolite, and DNA damage which forms the initiated cell. Through selective clonal expansion, this subsequently forms the preneoplastic lesion. Genetic change then occurs in the lesion, producing tumor conversion in which benign tumors are converted to malignant tumors (Belinsky et al. 1987; Harris 1985; Swenberg et al. 1985). In the initiation process, the replication of DNA with covalently bound adducts produces mutations in the daughter strand that are the heritable and irreversible changes involved in the initiation process. Promotion is cell proliferation leading to the selective clonal expansion of the initiated cell population. In the final progression stage of chemical carcinogenesis, cell proliferation is required for the irreversible progression from benign to malignant and metastatic neoplasms (Anderson et al. 1992).

The formation of adducts that initiate tumor formation does not always result in initiation since the damaged template can be repaired before cell replication has occurred. The role of tumor-suppression genes and oncogenes in chemical carcinogenesis has been discussed in Sec. II.A. It is believed that activation of proto-oncogenes and inactivation of tumor-suppression genes are required for expression of the tumorigenic phenotype. Certain PACs are known to induce the types of mutations that can activate proto-oncogenes to transforming oncogenes and inactivate tumor-suppression genes.

These mutations can produce structural changes in encoded gene products and loss of control of expression of these genes. For instance, Bowden and Krieg (1991) have found that mutagenic activation of the Harvey-*ras* proto-oncogene has been found to be associated with the initiation of mouse skin tumors by 7,12-dimethylbenz[*a*]anthracene. This activation was shown to be a result of a specific A : T transversion mutation at the second nucleotide of codon 61 of the Harvey-*ras* gene. The point mutations that cause activation of the oncogenes thus occur in certain "hot spots" on the genes, and if the sequence of specificity for the binding of carcinogenic PACs to DNA corresponds to a known biological hot spot in an oncogene, then the particular PAC can be a potent carcinogen (Stowers et al. 1987).

It has now been found (Hruszkewcyz et al. 1992) that PAH DNA adducts can cause inhibition of DNA polymerase activity and that *cis* adducts cause more inhibition than *trans* adducts. Thus, replication to form mutations should be less efficient with *cis* adducts.

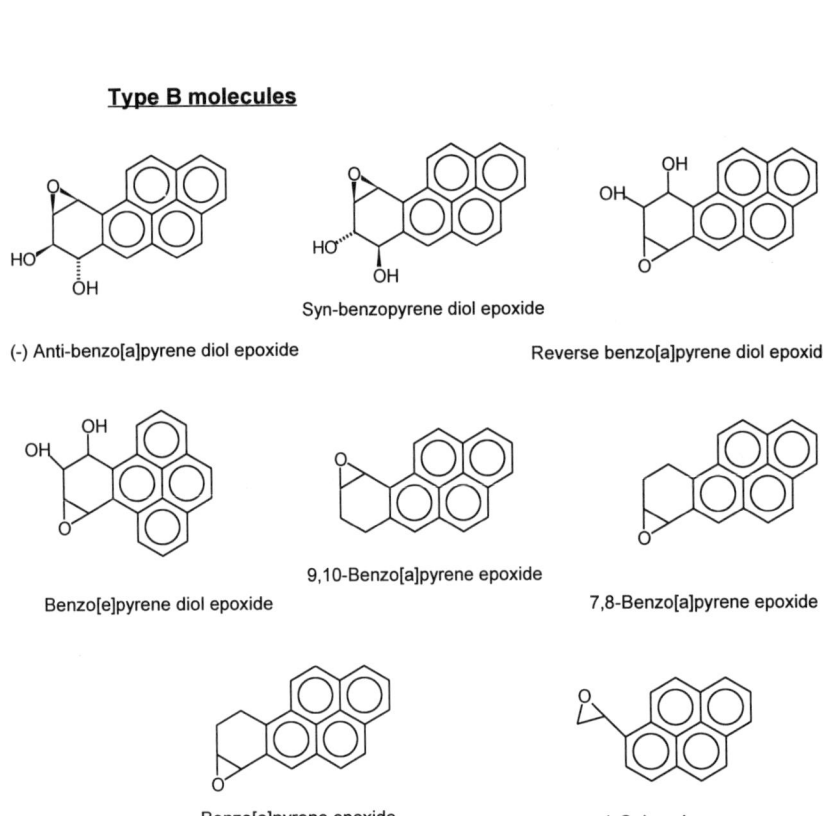

Fig. 10. Structures of various PAH epoxides and diol-epoxides. Type A are tumorigenic with change in conformation upon covalent binding, whereas type B are biologically less active with little change in conformation upon covalent binding.

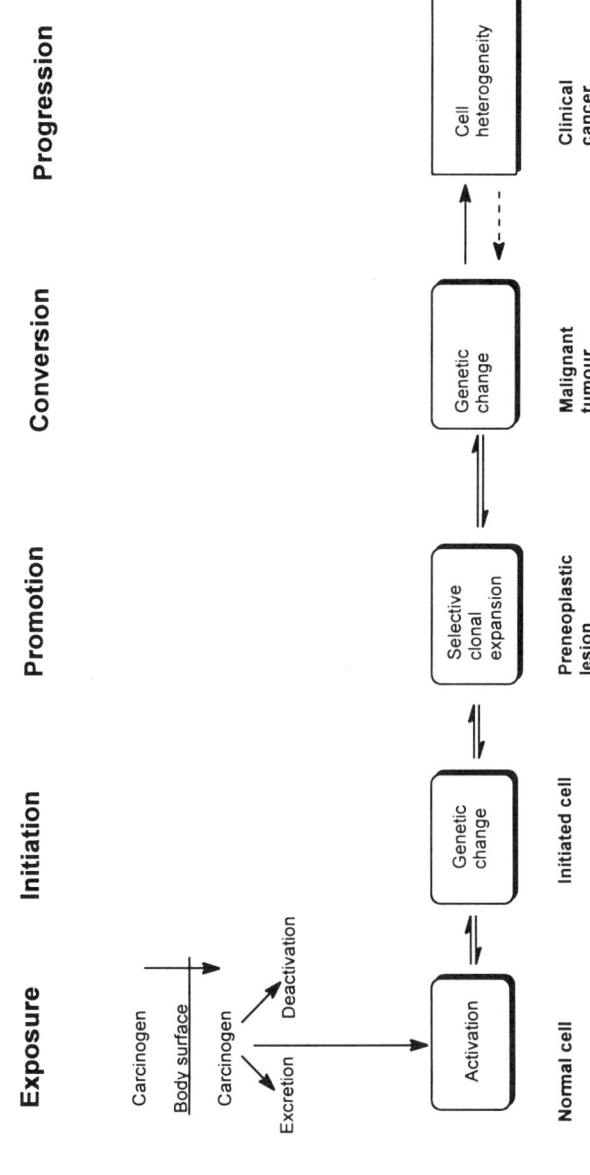

Fig. 11. Schematic representation of the multistage process of carcinogenesis in humans (Harris 1985).

The activity of carcinogenic PAHs has been found to be reduced in the presence of complex mixtures of PACs such as coal tar (Haugen and Peak 1983; Springer et al. 1986). This is believed to be due to suppression of DNA adduct formation by the carcinogenic PAHs caused by mass action with regard to binding sites on DNA and activating enzymes (Springer et al. 1986). Other studies, however, have found that the carcinogenicity of PAHs can be enhanced by the presence of soot particles (Pott and Heinrich 1990). Overall, the carcinogenicity of complex mixtures of PACs is poorly understood (IARC 1990).

III. Biomonitoring for Carcinogenicity
A. Use of DNA Adducts Formed in Human Tissues

1. Relevance of DNA Adduct Monitoring to Risk Assessment. The assessment of human health effects associated with exposure to chemical carcinogens is normally performed in four stages: (1) hazard identification, (2) exposure assessment, (3) dose-response assessment, and (4) risk characterization (Andersen et al. 1992; National Academy of Sciences 1983).

Risk characterization is achieved by combining exposure assessment and dose-response assessment. To enable the dose-response assessment to be undertaken most effectively, the "target dose" of a carcinogen should be estimated. The target dose has been defined as "the dose, expressed as time integral concentration, of the ultimate genotoxic agents (which may be systematically generated) which evades metabolic detoxification and penetrates to the biologically significant site in DNA" (ECETOC 1989; Farmer et al. 1987). Wide interindividual differences in PAC metabolism result in different DNA adduct levels; hence, measurements of external dose do not give an indication of target dose (Shields et al. 1992).

The benefit of determining target dose using DNA adducts in risk assessment is that the data obtained have a better predictive value than external or internal dose monitoring since modification of DNA is an important step in chemical carcinogenesis (ECETOC 1990; Montesano 1990; Wogan and Gorelick 1985). In addition, the association of specific DNA adducts with the activation of certain cellular oncogenes provides a link between the biochemistry of DNA damage produced by carcinogenic PACs and the molecular biology of carcinogenesis (Montesano 1990; Perera et al. 1990). The monitoring of adducts is thus regarded as the most applicable technique for risk assessment (Phillips and Hewer 1993). This is shown schematically in Fig. 12, where relevance to risk assessment increases from external dose measurements to the measurements of critical DNA adducts in target tissues.

The relationship between DNA adducts and carcinogenicity has been described by Lutz (1979) in terms of the covalent binding index (CBI), where

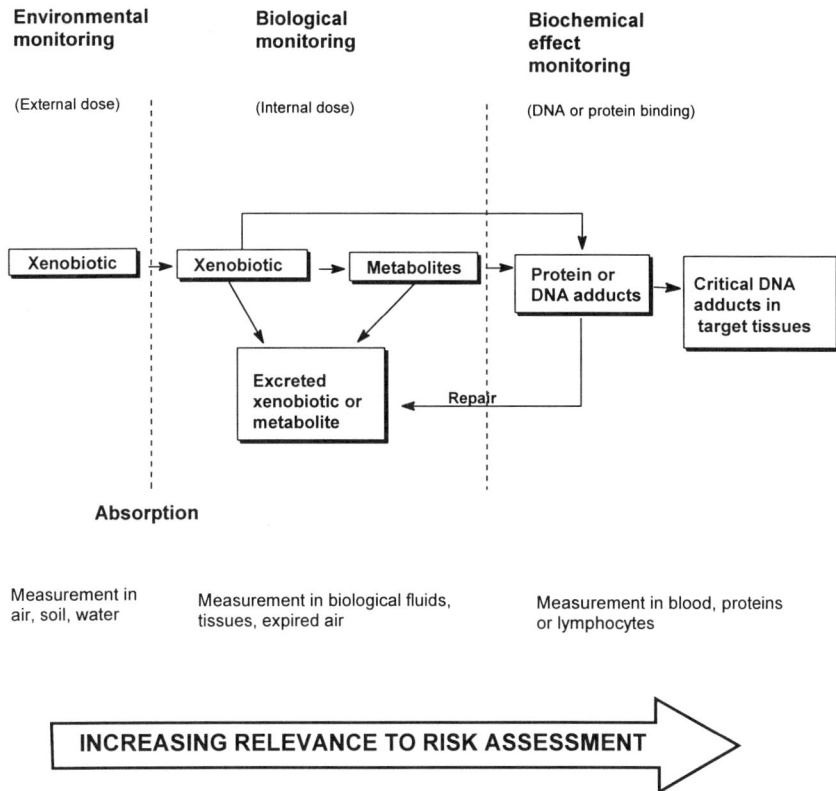

Fig. 12. Relevance of monitoring to risk assessment (ECETOC 1989).

$$CBI = \frac{\text{damage to DNA}}{\text{dose}}$$

$$= \frac{\mu\text{mol of chemical bound/mol of nucleotides}}{\text{mmol dose/kg body weight}}$$

The use of the CBI as a quantitative index of chemical carcinogenicity has been described as limited (Wogan and Gorelick 1985). However, Taningher et al. (1990) have found that the capability of the CBI to predict the initiating potential of chemical carcinogens (including PACs) was similar to many short-term tests such as mutagenicity, DNA damage/repair, and cell transformation tests.

Although the use of DNA adducts for biological effect (e.g., tumor initiation) monitoring has not been validated, biochemical effect monitoring using adducts has been used to monitor the extent of occupational exposure to various PACs. For instance, the monitoring of aromatic DNA adducts has been used to assess exposure to complex mixtures of PACs (Hemminki 1992), including coke by workers at coke ovens (Hemminki et al. 1990; Ovrebo et al. 1990), foundry workers (Hemminki et al. 1991), other coal combustion emissions (Mumford et al. 1993), and roofers exposed to coal tar pitch (Herbert et al. 1990). In addition, studies of occupational exposure to PAHs have used the specific detection of benzo[a]pyrene diol-epoxide (BPDE)-DNA adducts (Ovrebo et al. 1990; Vahakangas and Yrjanheikki 1990).

The limitations of the use of DNA adduct monitoring for risk assessment of exposure to carcinogenic PACs are related to the fact that DNA adducts are removed with different efficiencies in different genes in the same cell and that not all adducts cause the formation of tumors (Au 1991; ECETOC 1989). In addition, the quantitative predictability of tumor incidence by adduct concentration varies significantly across chemical classes and species, and more importantly, between species (Gaylor et al. 1992). Biochemical effect monitoring by DNA adduct determination is a rapidly developing field of research, and it has been recently stated (ECETOC 1989) that "a deeper insight into the mechanism of carcinogenesis is required before the measurement of DNA and haemoglobin adducts can be used confidently for human risk assessment—this should not preclude their use in exposure assessment." Research (Perera et al. 1992) has shown a dose-response relationship between PAHs and human cancers. In Silesia, Poland, a heavily polluted part of Europe, Perera et al. (1992) found that aromatic DNA adducts determined by the ^{32}P-postlabeling assay were significantly correlated with chromosome mutation, thus showing a molecular epidemiological correlation between exposure to PAHs and a genetic alteration relevant to cancer. In addition, the level of DNA adducts in gastric tumors was found to be significantly higher in smokers than nonsmokers, supporting epidemiological data that smoking is a high risk factor for gastric cancer (Dyke et al. 1992). DNA adducts have now been used as a measure of lung cancer risk of humans exposed to PAHs (Kriek et al. 1993).

2. Techniques for the Determination of DNA Adducts. The most sensitive method currently available for detection of PAC-DNA adducts is the ^{32}P-postlabeling assay (Beach and Gupta 1992; Friedberg and Hanawalt 1988; Gupta and Randerath 1988; Phillips 1990, 1992; Randerath and Randerath 1991). The procedure basically involves the following steps:

1. Enzymatic digestion of adducted DNA to deoxyribonucleoside 3'-monophosphates using micrococcal nuclease and spleen phosphodiesterase
2. Incubation of the digest from (1.) with [γ^{32}P]ATP in the presence of T4 polynucleotide kinase to form [5'-^{32}P]-deoxyribonucleoside-3',5'-bisphosphates (Fig. 13)
3. Separation of labeled adducted nucleotides from labeled unadducted nu-

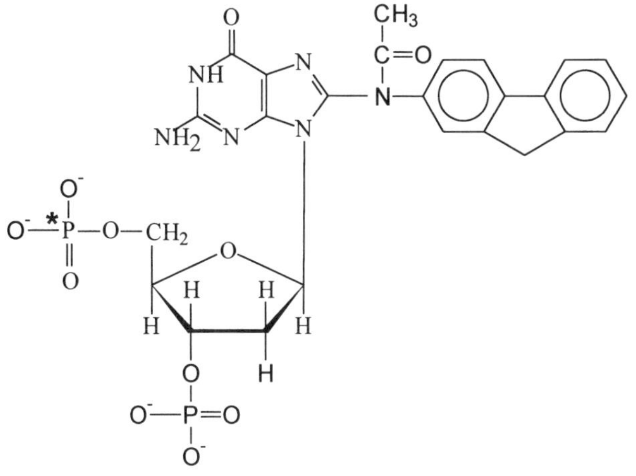

Fig. 13. 2-Aminofluorene adduct of [5'-^{32}P]deoxyribonucleoside-3,5,-bisphosphate.

cleotides by polyethyleneimine-cellulose TLC and detection by autoradiography
4. Quantification by scintillation counting

Various modifications to the procedure have been developed to allow for greater sensitivity (Lau and Baird 1992; Phillips 1990; Randerath et al. 1985; Randerath and Randerath 1991; Reddy and Randerath 1987; Vaca et al. 1992; Whong et al. 1992). In the standard procedure, the digested DNA (approximately 200 ng) is incubated with [γ-^{32}P]ATP (50–150μCi) and carrier ATP to give a slight molar recess of ATP. The use of a lesser amount of high specific activity (γ-^{32}P]ATP (>6000 Ci/mmol) instead of the usual molar excess of lower specific activity [γ-^{32}P]ATP results in the preferential labeling of adducted nucleotides compared to normal nucleotides. This can produce an increase in sensitivity of up to 50-fold for the same amount of radioactivity used. Increase in sensitivity is also increased by postincubation of DNA digests with nuclease P$_1$ before ^{32}P-labeling, which causes dephosphorylation of the normal nucleotides but not most PAC adducts, so that only the adducted nucleotides act as substrates for the labeling process. Enrichment can also be performed by extraction of adducted nucleotides into an immiscible organic solvent such as n-butanol before labeling. The unmodified ^{32}P-postlabeling technique can typically detect one adduct in 10^7 DNA nucleotides, whereas the modified techniques can detect one PAC adduct in 10^9–10^{10} DNA nucleotides. Better resolution of adducted nucleotides can be achieved using HPLC separation (Gorelick and Reeder 1993; Lecoq et al. 1992; Moller and Zeisig 1993).

The ^{32}P-postlabeling technique has been applied successfully to the estimation of DNA adducts formed by PAHs, heterocyclic PACs, nitroaromatics, and arylamines in vitro; in experimental animals in vivo; in wild animals; and in humans (Beach and Gupta 1992; Dyke et al. 1992; Gupta 1988; Herbert et al. 1990; Phillips et al. 1990; Mullaart et al. 1989; Roy et al. 1989; Varanasi et al. 1989; Weyand et al. 1993; Wolff et al. 1989). In addition to the detection of adducts formed by the covalent bonding of chemicals to DNA, the postlabeling assay can detect indirect modifications to DNA caused by chemicals that induce the formation of adducts by endogenous compounds (Phillips 1992). It should be noted that the ^{32}P-postlabeling assay will also detect uncharacterized modified nucleotides termed I compounds that arise indigenously without exposure to exogenous genotoxins (Randerath et al. 1989, 1990, 1993).

A number of other techniques have been developed and used in adduct determinations. These are discussed below:

(a) Postlabeling Assay Using [^{35}S] Phosphorothioate. This assay is based on the ^{32}P-postlabeling assay, but the adducted nucleoside-3'-phosphates are 5'-thiophosphorylated by T4 polynucleotide kinase and adenosine 5'-o-(3-[^{35}S] thiotriphosphate) to yield [^{35}S]-PAC-nucleoside-5'-phosphorothioate-3'-phosphate adducts (Lau and Baird 1991). In this method, unlabeled 3'-phosphates are selectively removed from the [^{35}S]-PAC-nucleoside-5'-phosphorothioate- 3'-phosphate adducts by a brief treatment with alkaline phosphatase. The separation of the nucleoside phosphorothioate adducts is achieved by high-performance liquid chromatography (HPLC).

(b) Immunochemical Methods. Polyclonal and monoclonal antibodies have been developed for the detection of DNA adducts (ECETOC 1989; Lau and Baird 1992; Phillips 1990; Santella et al. 1988, 1991). Methods used include competitive radioimmunoassays, enzyme-linked immunoassays (ELISA), and ultrasensitive enzyme radioimmunoassays (USERIA). These techniques have relatively high sensitivity and specificity, but the structure of analyzed adducts must first be determined since the corresponding hapten must be synthesized and coupled to a carrier protein and then the antibodies selected and characterized for sensitivity and specificity. The method is therefore restricted to the analysis of known adducts.

Immunochemical methods have been applied to the measurement of PAC-DNA adducts in both animal and tissue culture studies (Santella et al. 1984, 1985; Tierney et al. 1986). Several studies have monitored human exposure to PACs by the use of antibodies raised against benzo[*a*]pyrene diol-epoxide-DNA adducts. These have included the estimation of white blood cell DNA adducts in coke oven workers (Harris et al. 1985; Haugen et al. 1986) and iron foundry workers (Perera et al. 1988).

(c) Fluorescence Spectroscopy. Most PACs possess a distinctive fluorescence spectrum, but the fluorescence spectra of PAC-DNA adducts are often broad and highly quenched (ECETOC 1989; Phillips 1990; Sanders et al. 1986). The PAC-DNA adducts are therefore usually hydrolyzed in dilute acid to yield isomeric tetrols that have similar fluorescence intensities to the parent PACs (Weston and Bowman 1991). The DNA hydrolysates are then separated by reverse-phase HPLC and the fluorescence is continuously monitored at optimum excitation and emission wavelengths (Santella 1991).

The emission spectra are often complex, with many peaks. The synchronous fluorescence technique, however, scans excitation and fluorescence simultaneously with a fixed wavelength difference. A signal is only observed when the wavelength difference equals the interval between one absorption and one emission band, and in most cases a single peak is observed for each adduct (Johnson and Vo-Dinh 1989; Manchester et al. 1990; Phillips 1990; Weston and Bowman 1991). Because of this aspect of the assay, synchronous fluorescence spectroscopy (SFS) is applicable to the analysis of mixtures of adducts, especially when second-derivative synchronous fluorescence spectroscopy is used in combination with HPLC separation (Weston et al. 1989a). Recently, laser-induced fluorescence detection has been used with HPLC for detection of isolated PAH tetrols (Wang and O'Laughlin 1992). The lowest limit of detection achievable with these techniques is about one adduct in 10^8 nucleotides (ECETOC 1989; Weston and Bowman 1991).

Applications of SFS have included the determination of PAC-DNA adducts in coke oven workers (Haugen et al. 1986; Vahakangas and Yrjanheikki 1990; Weston et al. 1989b), human placenta (Weston et al. 1989a), human lung (Weston and Bowman 1991), and the alveolar macrophages of smokers (Izzotti et al. 1991).

(d) Gas Chromatography-Mass Spectrometry (GCMS). In order to determine DNA adducts by GCMS, the DNA must be hydrolyzed to release base adducts that are then concentrated by solvent extraction (Bolt et al. 1988; Lau and Baird 1992; Minnetian et al. 1987). Negative ion and fast atom bombardment MS appear to offer promise for DNA adduct work (Dino et al. 1987; Phillips 1990; Santella 1991), while tandem mass-spectrometric analysis has shown promise in the identification and quantification of unknown adducts (Cushnir et al. 1991; Farmer et al. 1993). The use of HPLC-MS with a thermospray or electrospray interface, however, permits the analysis of adducted nucleotides and has the advantage of providing information on their structure (Herreno-Saenz 1993).

B. Use of Protein Adducts Formed in Human Tissues

1. Relevance of Protein Adduct Monitoring to Risk Assessment. Hemoglobin and other protein adducts have been used as indicators for DNA adduct formation in target tissues such as liver, lymphatic, and lung tissues. The main reasons for the use of protein adducts in biochemical effect monitoring are their relatively easy accessibility compared with DNA adducts of target tissues and their relative stability (ECETOC 1989; Tornqvist and Kautianen 1993). Because of the stability of protein adducts in comparsion with DNA adducts, which constantly undergo repair, the concentration of protein adducts tends to represent exposure over life of the tissue monitored (e.g., hemoglobin adducts in red blood cells of man) and has been found to be proportional to the administered dose over a wide concentration range (Hemminki 1992; Neumann 1988; Shugart 1985; Skipper et al. 1984).

The feasibility of using protein adducts as an indicator of human exposure to PACs has been demonstrated by Lee et al. (1991), who showed correlations between serum albumin levels and occupational exposure to PACs by foundry workers and roofers. In addition, it has been found that 4-aminobiphenyl adducts are positively correlated to cigarette smoking, whereas PAH-DNA adducts have highly variable background levels (Perera et al. 1987, 1990). An important practical advantage of using protein adducts is that a better "response-to-background ratio" can be obtained in comparison with DNA adduct assays because relatively large quantities of protein (especially hemoglobin) can be obtained (Lohman 1988; Tornqvist and Kautianen 1993).

2. Techniques for the Determination of Protein Adducts

Immunochemical Methods. In a manner similar to DNA adduct determination, monoclonal and polyclonal antibodies have been developed for hemoglobin adduct determination. Sensitive quantification of PAC-protein adducts can be difficult due to low antibody affinity for the adduct, probably caused by shielding of the adduct in hydrophobic regions of the protein (ECETOC 1989; Santella et al. 1986). A four-fold increase in sensitivity with benzo[*a*]pyrene adduct determination, however, has been achieved by enzymatic digestion of proteins to peptides and amino acids before ELISA was performed (Lee et al. 1991).

Fluorescence Spectroscopy. This technique is suitable for the determination of adducts formed by strongly fluorescent molecules such as many PACs. The adducts must be hydrolyzed to form free tetrols which are then separated and detected fluorochromatographically by HPLC (ECETOC 1989; Haugen and Myers 1990).

Gas Chromatography and Gas Chromatography-Mass Spectrometry. Adducts on proteins must be hydrolyzed and extracted for GC and GC-MS analysis. Soft ionization techniques have recently found considerable use in the determination of protein adducts, including aromatic amines and PAHs. These techniques include fast atom bombardment (FAB), high-resolution selective ion monitoring, and tandem mass spectrometry (MS-

MS) (Farmer et al. 1988, Farmer et al. 1993; Skipper et al. 1984; Stilwell et al. 1987). The obvious benefit of GC-MS techniques is their ability to provide unequivocal identification of the adducts.

C. Comparison of the Application of DNA Adduct and Protein Adduct Monitoring

A consideration of the application of DNA and protein adduct determination techniques for measuring exposure and assessing health effects is presented in Table 3.

Limits of detection and quantities of samples required for DNA adducts are given in Table 4. The limits of detection for protein adducts of PACs by the methods listed above are generally in the order of 10 fmol adducts per gram of protein.

IV. Quantitative Structure Activity Relationships (QSARs)

The use of QSARs for the prediction of carcinogenicity of PACs has developed from a mechanistic approach in which descriptors are selected on the basis of their influence on the mechanism of DNA adduct formation. Alternative approaches have been based on structural relationships to activity of physicochemical properties often used with QSARs.

A. Mechanistic Approach

1. Unsubstituted PAHs. The most notable work involving this approach was the development of the bay-region theory (Jerina et al. 1976, 1977; Jerina and Lehr 1977). It was found that bay-region diol-epoxides produced the highest π orbital delocalization energy for ring opening of all diol-epoxides for a specific PAH. From this information, it is possible to rank relative carbocation delocalization energies for many different diol-epoxides. This process is represented in Fig. 14.

Jerina et al. (1976) calculated the change in π electron energy resulting from the additional aromatic conjugation ΔE_{deloc} by using the perturbational molecular orbital (PMO) methods of Dewar (1969). The ΔE_{deloc} value is dependent on changes in the coulomb integral α and change in the resonance integral β. In the PMO method, α represents the binding energy of an electron in its atomic orbital, and β the contribution to the π energy by electrons occupying the overlap region between atomic orbitals (see Dewar 1969).

A plot of calculated values of $\Delta E_{deloc}/\beta$ for various PAH carbonium ions by Jerina et al. (1977) is shown in Fig. 14. Higher values of $\Delta E_{deloc}/\beta$ indicate increased ease of carbonium ion formation. It can be seen from Fig. 14 that for benzo[a]anthracene, the bay-region carbonium ion is formed more easily than the nonbay-region ions. The right-hand portion of

Table 3. Evaluation of the Application of Molecular Methods for Measuring Exposure to Genotoxins[a]

	Method of Biological Monitoring			
	DNA adducts			
Criterion	Physical methods	Immuno-chemical methods	^{32}P-post-labeling	Protein adducts
Appropriateness for measuring exposure				
Qualitative	(+)	+	+	+
Recent (1 wk) internal dose	?	+	+	+
Long-term body burden	?	(+)	(+)	(+)
Dose at target site	?	+	+	−
Appropriateness for assessing health effects				
Nonadverse (reversible)	?	(−)	(−)	(−)
Adverse		(−)	(−)	(−)
Interpretation of results				
Individual basis	+	+	+	+
Group basis	+	+	+	+
Precision of method				
Technical reproducibility	?	(+)	(+)	+
Stability of parameter over time	(+)	(+)	+	(+)
Interlaboratory reproducibility	?	(+)	(+)	+
Sensitivity				
For environmental exposures	?	+	+	+
For occupational exposures	(+)	+	+	+
For acute exposures	+	+	+	+
Chemical specificity	+	±	−	±
Absence of interference by confounding factors	?	(+)	(+)	(+)
Absence of background levels	?	(−)	(−)	(−)
Simplicity of analysis	−	±	±	±
Ease of sample storage	+	+	+	+
Current applicability				
Research level	(+)	+	+	+
Routine use	(−)	(+)	(+)	(+)

[a] +, applicable/true; (+), probably applicable/probably true; −, not applicable/not true; (−), not presently applicable/not presently true; ±, cannot be generalized; ?, unknown.
Lohman (1988).

Table 4. Sensitivity of DNA Adduct Detection Methods

Method	Limit of Detection (fmol)	μg DNA Required	Adduct/Nucleotide
^{32}P-postlabeling	0.01	1–10	$1:3 \times 10^9$
Immunoassay			
RIA	40	to 10,000	$1:10^8$
ELISA competitive	1	50	$1:6 \times 10^8$
ELISA noncompetitive	3	0.1	$1:10^7$
Slot-blot	1	1	$1:3 \times 10^6$
Fluorescence			
Low-temperature	3	1000	$1:3 \times 10^8$
Synchronous scanning	20	100	$1:5 \times 10^6$
Line narrowing	1	1000	$1:10^8$
GC-MS	0.5	—	—

Phillips (1990).

the figure ranks various PAHs by their ease of bay-region carbonium ion formation.

The use of nonbonding molecular orbital (NBMO) coefficients A_{or} by Dipple et al. (1968), however, was the first attempt to relate carcinogenic activity to PAH metabolites. A delocalization coefficient 1-A_{or} based on the NBMO coefficient at sites of carbonium ion formation was used to estimate the relative stabilities of putative reactive metabolites. The index $\Delta E_{deloc}/\beta$ of Jerina et al. (1976) is equivalent to 2(1-A_{or}), where A_{or} is the NBMO coefficient for the atom carrying the positive charge (Dipple et al. 1984) (Figs. 15 and 16).

Since the original work by Dipple et al. (1968) and bay-region theory by Jerina et al. (1976), many approaches have been used to estimate carcinogenic potencies of PACs. Loew et al. (1985) have summarized the capabilities of theoretical chemistry in relation to the calculation of molecular descriptors of putative toxic effects, including characterization of physical interactions of suitable toxic species with biological target molecules and of chemical-biochemical interactions producing DNA adducts. This information is also presented in Table 5. The application of some of these principles and techniques has found use in QSARs for the carcinogenicity of PACs.

Smith et al. (1978) have examined a total of 21 parameters generally based on reactivity of parent compounds and metabolites and derived by molecular orbital calculations. Several of the parameters were estimations of molecular size or solubility, such as the n-octanol/water partition coefficient. As expected, the properties of metabolic intermediates showed greater correlations with carcinogenicity than the parent compounds. A moderate correlation was reported between carcinogenicity and Q_b for diol-epoxide-carbonium ion conversion. The term Q_b is related to Jerina's ΔE_{deloc} and Dipple's A_{or} by the following equations.

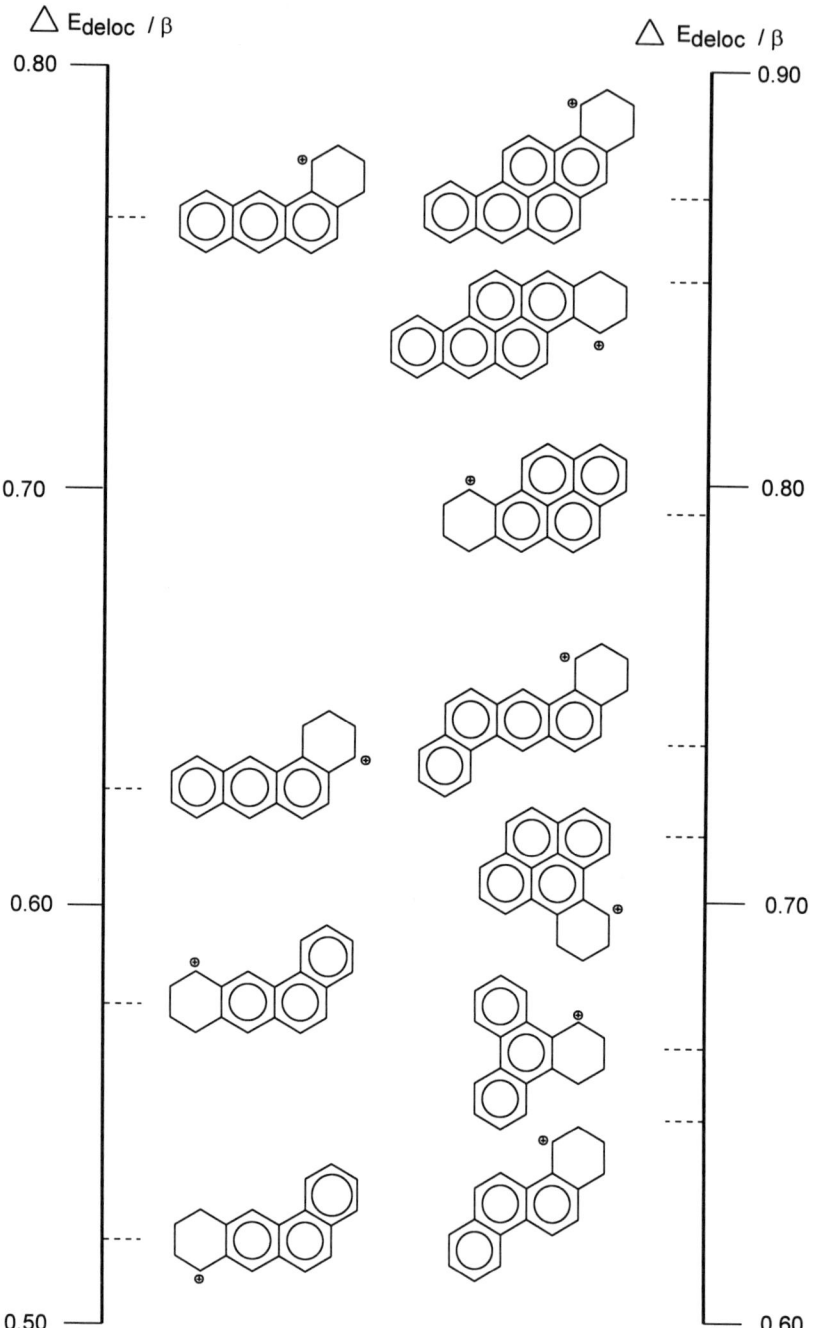

Fig. 14. Calculated values of $\Delta E_{\text{deloc}}/\beta$ for various PAH carbonium ions (Jerina et al. 1977).

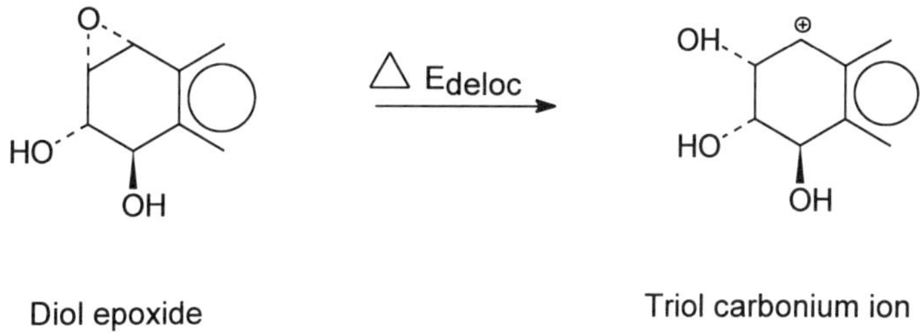

Fig. 15. Use of π electron delocalization energy to describe triol carbonium ion formation.

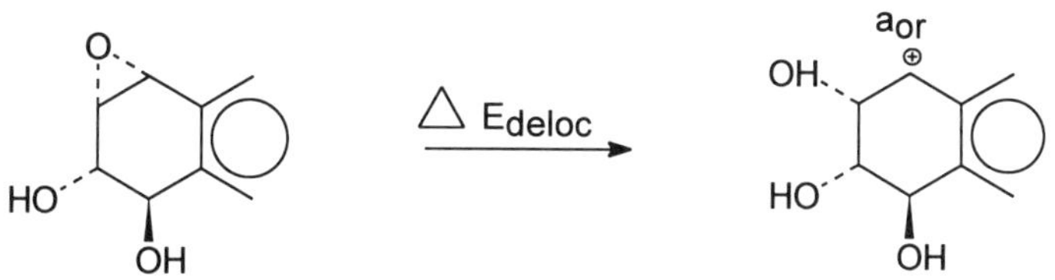

Fig. 16. Relationship between ΔE_{deloc} and a_{or} to describe triol-carbonium ion formation.

Table 5. Application of Theoretical Chemistry to the Calculation of Molecular Descriptors of Putative Toxic Effects

Calculated property	Application

Electronic properties
  Dipole and higher moments
  Ionization potentials
  Electron affinities
  Molecular electrostatic potentials
	Measure of the extent of complex formation with polar or nonpolar solvents and a variety of tissue macromolecules-DNA, proteins, membranes

Conformational energies Set of low-energy conformers	Measure of the nature and stability of complex formation with macromolecules
Chemical reactivity properties Group or atomic electrophilicities and/or nucleophilicities	Estimate of the relative ease and selectivity in transformation to specific intermediates and products Estimate of the extent and specificity of covalent adduct formation with tissue macromolecules

Loew et al. (1985).

$$\Delta E_{deloc}/\beta = 2(1 - Q_b)$$
$$A_{or} = Q_b$$

This work, however, has been criticized (Osborne 1979) for the use of Q_b that estimates the charge at the point of carbonium ion formation. Osborne (1979) has pointed out that the ease of carbonium ion formation within a series of related compounds depends mainly on the energy required to solvate the ion. This is estimated from the square of the charges at each carbon atom around the molecule $\Sigma |a_0|$.

In an attempt to overcome some of the limitations of the π-electron-only molecular orbital methods, semiempirical all-valence electron methods have been used to calculate reactivity because of their applicability to both planar and nonplanar molecules (Loew et al. 1979; Miertus et al. 1985). Positive correlations exist between carcinogenic potencies and (1) the reactivities of the parent PAHs for initial distal bay-region epoxidation and (2) the stabilities of the diol-epoxide carbocations.

Structure-resonance theory (RT) has been used to predict the carcinogenic activity of a series of PAHs (Herndon 1981). This model relates chemical reactivity to the difference in resonance energy between the PAH and cation intermediate. Resonance energy is calculated from the following equation:

$$\text{Resonance energy} = C \log SC$$

where C is a constant and SC (structure count) the number of principal resonance structures.

The above-mentioned techniques have been applied to the alternant PAHs, but the carcinogenic activity of a number of nonalternant cyclopenta containing PAHs is well known. The PMO methods used to determine the stability of carbocation intermediates formed from an electrophilic attack on alternant PAHs have been modified for application to the cyclopenta PAHs (Gold et al. 1988). Electrophilic attack at the carbon atoms of the etheno bridge produces an odd-alternant system that can be treated in a manner similar to that for alternant PAHs. Electrophilic attack at any carbon atom that leaves the etheno bridge intact, however, produces a nonalternant system containing a five-membered ring. With this type of system, the ease of electrophilic attack has been estimated (Dewar and Dougherty 1975; Gold et al. 1988) by uncoupling the bond of the cyclopenta ring to thereby form an alternant PAH. Thus, $|LE/\beta|$ is adjusted by the quantity $-2a_{ot}a_{ou}$, where a_{ot} and a_{ou} are the NBMO coefficients of the uncoupled carbon atoms, and the product $-2a_{ot} a_{ou}$ represents the decrease in resonance stabilization $\Delta E_\pi/\beta$ resulting from the recyclization of the cyclopenta ring at C_u and C_t (Fig. 17). The term $|LE/\beta|$ represents localization energy for the attack of principal carbon atoms with subsequent formation of carbocation intermediates. This PMO approximation, however, has the limitation that it does not account for steric interactions or geometric dis-

Fig. 17. Uncoupling of *t-u* bond in cyclopenta-PAH.

tortions and has failed to predict the activated metabolites of benz[l]aceanthrylene, which is highly strained and nonplanar (Gold et al. 1988).

In the development of QSARs for carcinogenicity of PAHs, solvation of the carbocation intermediate can also be taken into account (Jerina et al. 1982). Solvation energies of the carbocations can be calculated from the distribution of charges around the molecules, which are considered to be in a medium of given dielectric constant (Roussel et al. 1988). The solvation energy decreases as the charge distribution is extended, whereas the carbocation stability increases as π electron delocalization increases. The ease of formation of PAH carbocations can therefore be reduced from that calculated by PMO theory (ΔE_{deloc}) by a decrease in solvation energy for certain molecules. This can be observed for carbocations of the series of linearly fused PAHs shown in Fig.18.

It is apparent that many PAHs can be metabolically activated to proximate and ultimate carcinogens, but metabolism to a bay-region diol-epoxide is not sufficient to produce carcinogenic activity (Kapitulnik et al. 1978; Wislocki and Lee 1988). The bay-region theory that uses reactivity descriptors and other subsequently developed descriptors does not take into consideration stereochemical or conformational parameters that may influence the amount of metabolically activated intermediates that react with nucleophilic sites on DNA. Once the PAH has been activated to the diol-epoxide stage, steric and conformational parameters in addition to reactivity determine which are the ultimate carcinogens (Wislocki and Lee 1988). Studies involving the diol-epoxides of various PAHs, including benzo[a]pyrene, benz[a]anthracene, and chrysene have found that only the

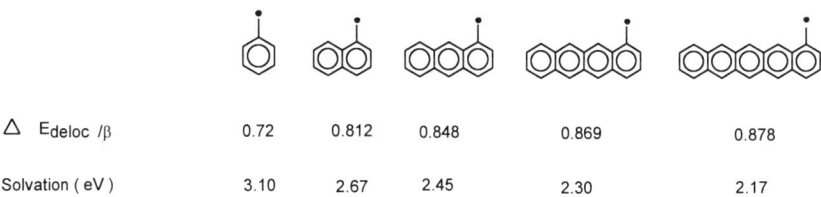

Δ E_{deloc}/β	0.72	0.812	0.848	0.869	0.878
Solvation (eV)	3.10	2.67	2.45	2.30	2.17

Fig. 18. Carbocation stabilities and solvation energies for linearly fused PAH carbocations (Roussel et al. 1988).

diol-epoxides in which the -OH groups are in the diequatorial conformation possess tumorigenic activity, in comparison with the diaxial-OH diol-epoxides, which possess little activity (Coles 1984; Jerina et al. 1984). This can be observed in a comparison of the tumorigenicity of the bay-region diol-epoxides of benzo[c]phenanthrene, which possess hydroxyl groups in the diequatorial conformation, and benzo[e]pyrene and triphenylene diol-epoxides, whose hydroxyl groups are present in the diaxial conformation (Agarwal et al. 1987; Wislocki and Lee 1988). The bay-region diol-epoxides of benzo[e]pyrene and triphenylene are both inactive, whereas both the bay-region diol-epoxides of benzo[c]phenanthrene are tumorigenic. The diaxial conformation of the hydroxyl groups may prevent the diol-epoxides from reacting at critical sites on DNA (Wislocki and Lee 1988). It appears the conformation of an adduct may be influenced by the base sequence at the point of adduct formation, and that the conformation of the adduct causes a different pattern of mutation upon replication (Rodriguez and Loechler 1993). It is also possible that the interaction of DNA repair enzymes with DNA may be differentially affected by the conformation of the adducts.

In addition to the fact that the correct hydroxyl group conformation of diol-epoxides is necessary for tumorgenic activity, the stereochemistry of the diol-epoxides also affects their tumorgenicity (Cody 1985; Harvey 1981; Wislocki and Lee 1988). This is shown in Table 6, which compares the relative tumorigenic activity of the two optical isomers of each diol-epoxide of several PAHs.

Covalent binding of the ultimate PAH carcinogens to DNA appears to be affected by the degree of physical binding of these diol-epoxides (Geactinov 1988; Harvey 1981). In addition to covalent binding of activated PAH metabolites to DNA, intercalation can occur as shown in Fig. 19. Intercalation of some diol-epoxides to DNA can enhance their covalent binding to DNA bases (Paulius et al. 1986; Prakash et al. 1988). Physical binding to other macromolecules such as proteins is believed to compete with covalent binding to DNA of PAH-activated metabolites. Geactinov (1988) has expressed this process by the equation below:

$$f_{cov} = \left[\frac{K_c}{K_c + K_T(1)}\right] \left[\frac{K_3(1)}{K_h}\right] K_1 \text{ (DNA)}$$

where f_{cov} is the fraction of diol-epoxide molecules undergoing covalent binding, K_c the assocation constant for formation of covalent adducts from carbonium ions, $K_{T(l)}$ the association constant for the formation of tetrols from carbonium ions, $K_{3(l)}$ the rate of formation of carbonium ions, K_h the hydrolysis rate constant for the diol-epoxide, and K_1 the association constant for physical binding.

2. Alkylated PAHs. The alkylation of PAHs can have a marked effect on their carcinogenicity. For instance, benz[a]anthracene, a weak carcinogen,

Table 6. Relative Tumorigenic Activity of Enantiometric Diol-epoxides of Several PAHs

Compunds	Initiation promotion model	Newborn mouse model
Benzo(a)pyrene		
(+)-diol-epoxide-1[a]	0	20
(−)-diol-epoxide-1	0	10
(+)-diol-epoxide-2[b]	90	900
(−)-diol-epoxide-2	0	0
Benz(a)anthracene		
(+)-diol-epoxide-1	12	3
(−)-diol-epoxide-1	3	0
(+)-diol-epoxide-2	60	270
(−)-diol-epoxide-2	2	0
Chrysene		
(+)-diol-epoxide-1	2	0
(−)-diol-epoxide-1	1	0
(+)-diol-epoxide-2	8	10
(−)-diol-epoxide-2	2	0
Benzo(c)phenanthrene		
(+)-diol-epoxide-1	600	30
(−)-diol-epoxide-1	80	20
(+)-diol-epoxide-2	300	500
(−)-diol-epoxide-2	1000	1000

[a]Diol-epoxide-1 indicates diaxial conformation.
[b]Diol-epoxide-2 indicates diequatorial conformation.
Wislocki and Lu (1988).

becomes a potent carcinogen when dimethylated in the 7- and 12-positions. On the other hand, alkylation can reduce the carcinogenicity of PAHs. For instance, methylation on the benzo ring of benzo[a]pyrene (positions 7, 8, 9, or 10) considerably lowers the potency of this carcinogen (Silverman and Lowe 1988). In addition, methylation in the "peri" position adjacent to the reactive benzo ring (e.g., position 12 in 5-methylchrysene or position 6 in benzo[a]pyrene) inactivates the PAHs by apparently inhibiting the metabolism to bay-region diol-epoxides (Cody 1985; Hecht et al. 1988; Loew et al. 1985; Silverman and Lowe 1988). Figure 20 shows the positions of peri methyl substitution causing deactivation for 5-methylchrysene and benzo[a]pyrene.

Methyl substitution that causes activation of PAHs normally occurs in the bay region (e.g., position 12 of benz[a]anthracene or position 11 in

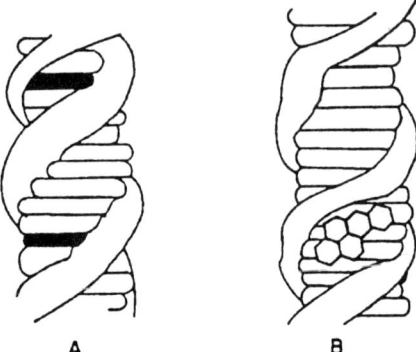

Fig. 19. Schematic representation of *anti*-BPDE (a) intercalated between the base pairs of the DNA helix and (b) covalently bound to DNA and residing in a minor groove (Harvey 1981).

benzo[*a*]pyrene). Activation, however, can occur by substitution at "remote sites" that are not on or adjacent to the reactive benzo ring (Silverman and Lowe 1988). An example of this is methylation at position 7 of benz[*a*]anthracene.

Carcinogenic activation of PAHs by methyl substitution, both in the bay region and at remote sites, has been explained by Silverman and Lowe (1988) in terms of lowest unoccupied π molecular orbital (LUMO) considerations. Since carbocation stabilization resulting from the diol-epoxide ring opening is achieved by transfer of electron density to the carbocation from the aromatic molecule, then electron-releasing substituents such as methyl groups can stabilize the carbocation by electron transfer from the methyl group to the empty aromatic delocalized levels of the PAH. The magnitude of the LUMO at various substitution sites determines the carbocation stabil-

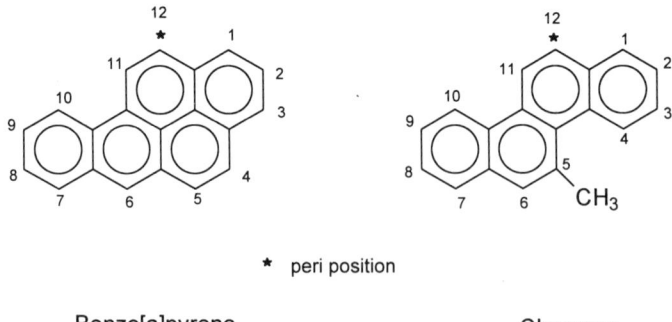

Fig. 20. Peri positions adjacent to the reactive benzo rings for benzo[*a*]pyrene and 5-methylchrysene.

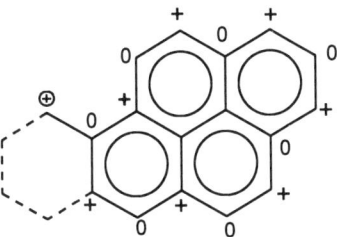

Fig. 21. Nonvanishing (+) and vanishing (0) LUMO amplitudes for benzo[a]pyrene bay-region carbocation.

ity when substitution occurs at these sites. The substitution sites have either vanishing or nonvanishing LUMO amplitudes that can be determined from molecular topology (Silvermann and Lowe 1982, 1988). Methyl substitution at sites of nonvanishing LUMO amplitude should stabilize the carbocation, whereas substitution at sites of vanishing LUMO amplitude should have little or no effect. Figure 21 represents the benzo[a]pyrene carbocation formed from opening of the bay-region diol-epoxide, together with nonvanishing and vanishing LUMO amplitudes marked by "x" or "o".

The enhancement of carcinogenicity by methyl substitution in the bay region has also been explained by the unique properties of these substituted bay-region dihydrodiol-epoxide metabolites (Hecht et al. 1988). It was postulated that the position of the methyl group adjacent to the bay region may be important in the alignment of the diol-epoxide for covalent binding to DNA. Studies using x-ray crystallography have shown that conformational differences exist between bay-region methylated and unsubstituted PAHs (Cody 1985; Hecht et al. 1988; Kashino et al. 1984; Zacharias et al. 1984). It appears that distortions both in-plane from the widening of the bay region and out-of-plane from torsion about the bay-region bonds are caused by steric interactions between methyl hydrogens and ring hydrogens, and that these distortions assist in the interactions between activated metabolite and DNA required for covalent adduct formation (Cody 1985; Hecht et al. 1988).

The inhibition of carcinogenicity of PAHs by methyl substitution in the peri position has been explained in terms of inhibition of formation of the dihydrodiol in the adjacent angular benzo ring (Amin et al. 1982; Hecht et al. 1988; Silverman and Lowe 1988; Yang et al. 1980). A second possible mechanism for the inhibiting effect of peri methyl substitution involves the conformation of the angular ring dihydrodiols (Hecht et al. 1988; Loew et al. 1985; Yang 1988). It is known that when the dihydrodiols possess hydroxyl groups in the diequational conformation, they are readily converted to the bay-region dihydrodiol-epoxides, whereas when the hydroxyl groups are present in the diaxial conformation, they are frequently not enzymatically converted to the corresponding dihydrodiol-epoxides (Hecht et al.

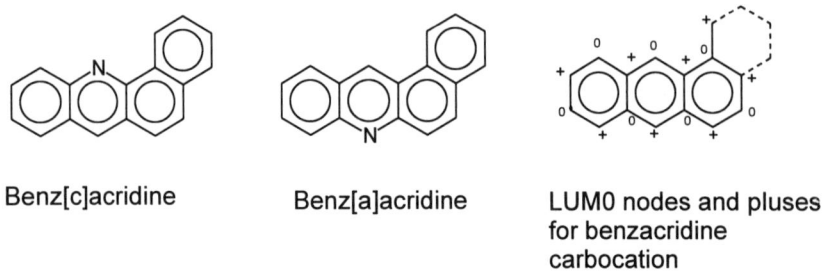

Fig. 22. Benzacridine structures and distribution of positive charge for the LUMO.

1988; Lehr et al. 1985). It is also known that dihydrodiols with a peri methyl substituent adopt quasidiaxial conformations (Yang 1988); thus, it appears that the conformation of the peri methyl-substituted dihydrodiols inhibits their conversion to bay-region dihydrodiol-epoxides.

3. Heterocyclic and Nitrogen Substituted PACs. The heterocyclic PACs that have been mostly studied in respect to QSARs for carcinogenicity are the azaarenes such as the benzacridines (Lehr et al. 1988; Silverman and Lowe 1988). These compounds have a carbon atom or carbon and hydrogen atoms replaced by nitrogen. Bay-region diol-epoxides are thought to be the ultimate carcinogenic metabolites of these aza PACs (Chang et al. 1984; Lehr et al. 1988; Levin et al. 1983). The reactivity of the diol-epoxides can be determined using HMO theory in a manner similar to that for PAHs.

The effect of ring nitrogen on carbocation stability depends on position of the ring nitrogen. The structures and LUMO nodes and pluses for benzacridines are shown in Fig. 22. Nitrogen atoms are electronegative in comparison with carbon atoms, and nitrogen substitution at points of nonvanishing LUMO charge will attract some of the charge used for stabilization of the carbocation, resulting in a reduction of the reactivity of the carbocation (Lehr et al. 1988; Silverman and Lowe 1988). In line with this, it has been found that benzacridine bay-region diol-epoxides are less active than benz[*a*]anthracene diol-epoxides (Silverman and Lowe 1988). In addition, benz[*a*]acridine is less active than benz[*c*]acridine, as would be expected since N-substitution on a plus LUMO point will destabilize the carbocation. It has been found (Lehr et al. 1988) that there is a correspondence between carbocation destabilization and the square of the LUMO amplitude at the various points of N-substitution with the exception of position 12 (Lowe and Silverman 1983). This is possibly caused by a decrease in steric crowding caused by removal of a hydrogen atom in the bay region (on carbon atom 12) with a resultant effect on carbocation stabilization (Silverman and Lowe 1988).

Nitro-PACs and amino-PACs will be discussed together since metabolic activation of both aryl-nitro and arylamine compounds can involve conversion to hydroxylamines, then esters, and subsequently aryl nitrenium ions (Debnath et al. 1991; Loew et al. 1985). In an analogy with bay-region PAHs, it would be expected that the ease of formation and stability of the electrophilic nitrenium ions would be determining factors in the carcinogenicity of arylamines and nitro-aromatics. In fact, ease of formation of nitrenium ions or their stability calculated as delocalization energies do not provide any consistent correlation with carcinogenic activity of the parent amine (Silverman and Lowe 1986). This may be due to selective metabolitic formation of reactive intermediates (Silverman and Lowe 1988).

Loew et al. (1985) have considered that the carcinogenicity of aromatic amines is influenced by activation processes such as nitrenium ion formation and to a certain extent ring epoxidation and detoxifying processes such as phenol formation. Loew et al. (1985) have proposed two properties to describe formation of arylnitrenium ions [$SN(\pi)$ and $\Delta E_{NH}+$], one to describe formation of inactive hydroxyl metabolites [$S_c^{max}(\pi)$], and two to account for adduct formation (ρ_N and $\rho c\beta$) in LUMO. The three processes involved in determining arylamine mutagenicity are shown in Fig. 23. The relative proportions of N-oxidation to arylnitrenium ions compared to ring oxidation to phenols have been correlated with measured half-wave potentials $E_{1/2}\Delta$ using linear sweep volumetry by Kadlubar et al. (1990)

B. Non-mechanistic Approach

1. Artificial Intelligence Systems: Structure Recognition. Computer-based systems, such as the computer-automated structure evaluation (CASE) program, automatically select the substructure units that are most appropriate to discriminate between active and inactive molecules (Klopman 1985). These fragments are recognized by taking linear portions of carcinogenic and noncarcinogenic molecules and evaluating each fragment in the computer database. If a fragment is randomly distributed in both the carcinogenic and noncarcinogenic database, it will be nonactive. On the other hand, a fragment is considered to contribute to carcinogenicity if its distribution is significantly different from random at a 95% confidence level (Klopman 1985; Klopman and Rosenkranz 1991; Rosenkranz 1992). In addition to the identification of active fragments (biophores), this program also identifies functionalities that contribute to a lack of activity (biophobes). A list of compounds consisting mainly of PACs containing a common biophore is shown in Table 7. It can be observed, however, that this biophore is active with respect to mutagenicity in *Salmonella*, but not necessarily with respect to carcinogenicity. Table 8 lists biophores and biophobes plus QSAR results for PAH carcinogenicity.

Of the three biophores (regression coefficient >1), biophore 1 is the most prevalent and represents a bay region. The most carcinogenically active biophore, however, is number 3 (regression coefficient 1.72). This biophore is shown in Fig. 24 and is structurally similar to a portion of the

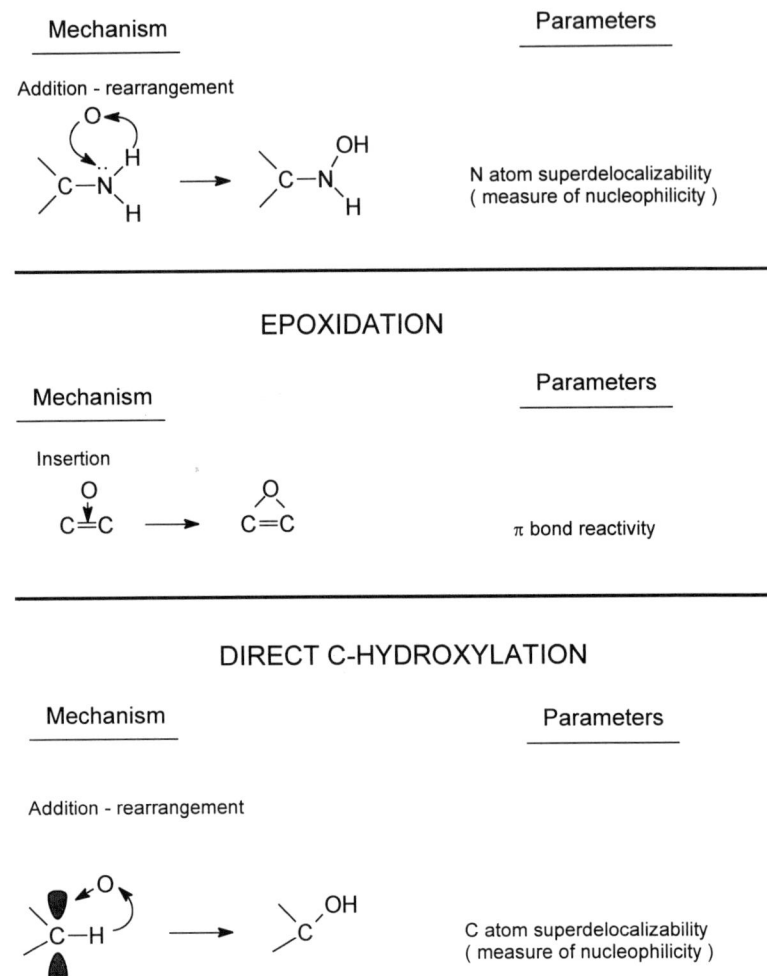

Fig. 23. Proposed mechanisms influencing carcinogencity of arylamines and parameters for their estimation (Loew et al. 1985).

benzo[*a*]pyrene molecule plus several other PAHs. Table 9 shows fragments from arylamines that may contribute to carcinogenicity in rodents. Figure 25 illustrates both activating and deactivating fragments with respect to the mutagenicity of nitro-PAH. It should be realized that the applicability of artificial intelligence systems in the prediction of carcinogenicity of PACs depends on the system's ability to evaluate relatively large molecular moieties and should be capable of incorporating possible mechanisms for carcin-

Table 7. Molecules Containing a Common Biophore That Is Significant with Respect to Mutagenicity in *Salmonella*

$$\begin{array}{c} CH=CH \\ \diagdown \\ CH \\ /\!/ \\ -C \\ \diagdown \\ CH \end{array}$$

o-phenanthroline	D and C yellow II
9-nitroanthracene	1-nitro-2-methylnaphthalene
CI pigment red 3	CI pigment red 23
D and C red 9	8-hydroxyquinoline
CI solvent yellow 14	N-phenyl-2-naphthylamine
benzo[a]pyrene	3-methylcholanthrene
benz[a]anthracene	phenanthrene
benzo[f]quinoline	1-nitronaphthalene
naphthalene	quinoline
2-naphthylamine	coumarin
anthracene	pyrene
1-naphthylamine	benzo[e]pyrene
1,8,9-trihydroxyanthracene	1-naphthylisothiocyanate
2-anthramine	7,9-dimethylbenz(c)acridine
N-(1-naphthyl)ethylenediamine	7-bromomethyl-12-methylbenz[a]anthracene

Klopman and Rosenkranz (1991).

ogenic activity. It also appears that these carcinogenicity prediction programs for PAHs primarily emphasize the importance of the bay region to carcinogenesis (Klopman 1985; Mitchell et al. 1986).

This procedure has been regarded by ECETOC (1986) as being limited since the descriptors are based only on structure recognition, and therefore the QSAR can only be used when the carcinogenic activity does not depend on physicochemical properties.

2. Physicochemical Descriptors. The use of physicochemical descriptors alone produces poor correlations with PAC carcinogenicity, and these descriptors are usually used in conjunction with formation reactivity of the ultimate carcinogens (Benigni and Guiliani 1987; Benigni et al. 1989; Debnath et al. 1991; Filov and Ivin 1985; Miertus et al. 1985; Szentpaly 1984). Physicochemical descriptors that have been used in QSARs for the carcinogenicity of PACs include those representing molecular size (number of carbon atoms, total surface area, total molecular volume, molecular weight, molecular refractivity, and connectivity indices) and those repre-

Table 8. Quantitative Structure-Activity Relationship Results for Carcinogenic Activity of Polycyclic Aromatic Hydrocarbons

Variable[a,b]	Regression Coefficient	Standard Deviation
	(Constant = 1.847)	0.204
1	1.074	0.243
2	1.159	0.336
3	1.721	0.505
4	−0.354	0.169
5	−0.728	0.334
Standard deviation of residuals		0.813
Index of determination		0.832

[a]Subunit 1: C·=C−CH=CH−CH=CH−C=CH−C·=CH−C·=
Subunit 2: C·=CH−CH=CH−CH=C·−C·=C·−CH=CH−
Subunit 3: C·=CH−CH=CH−C·=CH−CH=C·C·=CH−
Subunit 4: CH=C·−CH=CH−CH=CH−C·=CH−
Subunit 5: CH=CH−CH=C·−CH=CH−C·=CH−CH=CH−
DOT (·) indicates that the previous atom bridges two rings.
[b]No. of molecules = 43; no. of false negatives = 7, of which 3 are marginal; no. of false positives = 0.
Klopman (1985).

senting partition across polar/nonpolar interfaces (the *n*-octanol/water partition coefficient, K_{ow}).

Most enzyme receptor systems show an optimal size dependence for the receptor. The enzyme systems for the metabolic activation of PACs, however, must be capable of interacting with a broad range of compounds and consequently do not possess highly discriminating substrate specificities (Guenther and Oesch 1981; Szentpaly 1984). There does appear, however, to be an optimum size for PAHs to exhibit carcinogenic activity, and PAHs containing between 20 and 24 carbon atoms are the most potent (Szentpaly 1984).

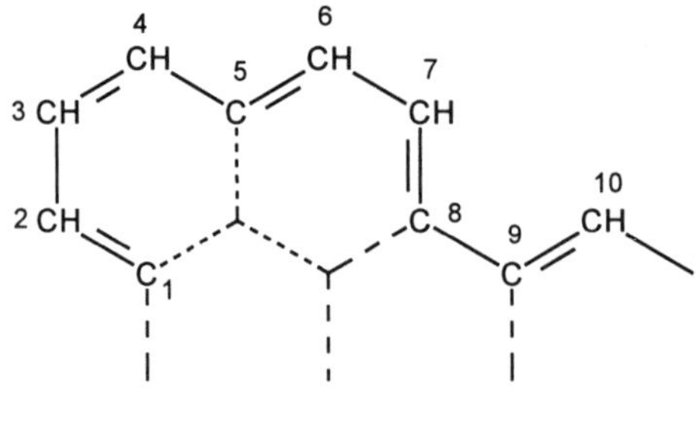

Fig. 24. Structure of biophore 3 (Klopman 1985).

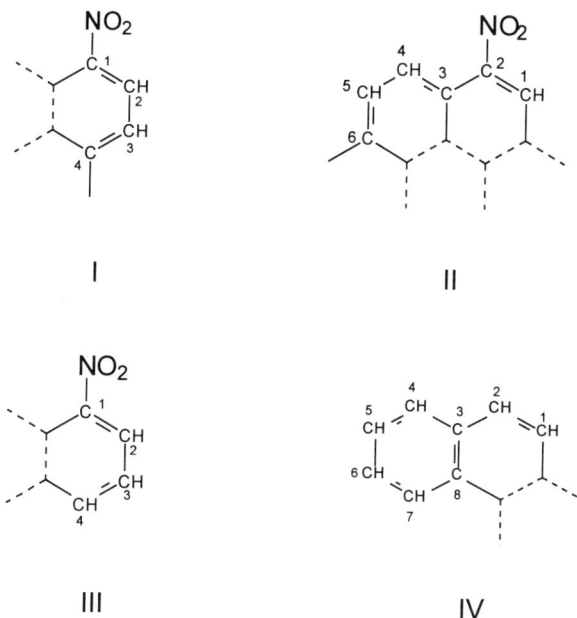

Fig. 25. Fragments responsible for the mutagenicity of nitro-PAHs; I or II are required for activity, whereas fragments III and IV are deactivating (Rosenkranz and Mermelstein 1985).

Table 9. Relationship Between Structure of Aromatic Amines and Carcinogenicity in Rodents

Fragment	Total	Inactive	Equivocal	Active	% Actives and equivocals
All molecules	252	88	29	135	65.1
$NH_2-C=$	53	12	5	36	77.4
$NH_2-C=CH-CH=$	38	10	5	24	76.3
$NH_2-C=C-CH=CH$[a]	19	3	0	16	84.2
$NH_2-C=CH-CH=C-C-NH_2$[a]	4	0	0	4	100

[a]CASE identified these aromatic amine-derived fragments as associated with an increased probability of carcinogenicity.

Klopman and Rosenkranz (1991).

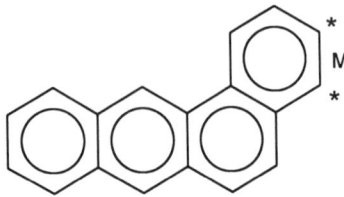

Fig. 26. Site of epoxidation for benz[a]anthracene.

In order for the PACs to be metabolically activated, they must reach the activating enzymes in the microsomal endomembrane system (Guenther and Oesch 1981; Sims and Grover 1981; Szentpaly 1984). A more lipophilic PAC will partition into the lipid phase of the membrane more easily and thus be metabolically activated to a greater extent. The more polar activated products such as dihydrodiol-epoxides and hydroxylamines, however, should be more hydrophilic to partition back into the aqueous medium surrounding the cell nucleus. It is unclear from the literature as to whether the lipophilicity of the parent PAC dominates the transport process or whether a combination of lipophilicity of the parent compounds and hydrophilicity of the metabolites is a determining factor in the carcinogenesis of PACs. If the latter case is true, then an optimum partition coefficient for the PACs may exist.

C. Combined Approach for QSARs to Predict Carcinogenicity of PACs

It is now very apparent that the use of a single descriptor will not accurately predict the carcinogenic potency of PACs. There is a strong dependence on the mechanisms of activation of PACs to their ultimate carcinogens such as the bay-region theory and to a lesser extent the distribution of the PACs to the sites of activation and DNA binding.

Examples of the use of combined approaches are those of Szentpaly (1984) and Miertus et al. (1985), who have correlated the carcinogenicity of PAHs with the ease of formation of the initial bay-region epoxides, ease of formation of the carbocation, and either the size of the parent PAH or its lipophilicity.

In this approach, epoxidation at the M region (Fig. 26) in competition with other areas of the PAH molecule is calculated. The ease of epoxidation in the M region has been found to be negatively correlated with the smaller of the two Dewar reactivity numbers N_m representing the carbon atoms marked with asterisks in Fig. 26 (Szentpaly 1984). The Dewar reactivity number is defined as

$$N_m = 2(C_{o,m-1} + C_{o,m+1})$$

where $C_{o,m-1}$ and $C_{o,m+1}$ are the NBMO coefficients at atoms $m - 1$ and $m + 1$ resulting if the π electron system is interrupted at atom M. A metabolic index for epoxidation of the M region is shown below.

$$M = (N_m - N_c)^2$$

where N_c is the smallest Dewar number corresponding to competitive reactions at other positions.

The ease of carbocation formation has been estimated by Szentpaly (1984) using the descriptor $(E_D + E_C)$, where E_D is the Huckel delocalization energy for the delocalized carbocation and E_C the charge dispersal energy used to correct for the fact that the Huckel term does not distinguish between the formation of a cation and a radical. The importance of the inclusion of the E_C term is observed in a comparison of the correlation coefficients for the equations below:

$$I = -146.33M + 38.91 E_D - 0.0569\Delta - 614.17 \ (n = 26, r = 0.824, S = \pm 14.1)$$

$$I = -80.47M + 8.244(E_D + E_C) - 0.073\Delta - 331.7 \ (N = 26, r = 0.961, S = \pm 6.8)$$

where I is the Iball index for carcinogenicity and $\Delta = (n - 20)^3$, with n being the number of carbon atoms in the PAH. For the 26 PAHs examined, this QSAR (previous equation) produced an excellent correlation coefficient, and with the E_C term it does not give rise to any false positives. This approach has been described in an assessment of QSARs (ECETOC 1986) as using appropriate descriptors and being a valid QSAR for the carcinogenicity of PAHs.

V. Conclusions

A wide range of carcinogenic activities occur in the PACs. These compounds must be metabolically activated to their ultimate carcinogens by cytochrome-P450 and other enzymes. The PAHs are activated predominately by formation of bay-region diol-epoxides and subsequently reactive carbonium ions. Aromatic nitro- and amino-compounds are activated by the formation of hydroxylamines and subsequently reactive nitrenium ions although bay-region diol-epoxides may also be formed.

Cancer formation is a complex process that is seen as initially involving the formation of adducts on DNA. It is postulated that the formation of DNA adducts may be responsible for the activation of oncogenes and deactivation of tumor-suppression genes. The stereochemistry of activated metabolites is important with respect to DNA adduct formation as only certain enantiomers (usually in the anticonformation) cause tumor formation.

The determination of DNA adducts of PACs can be used as a biochemical effect monitoring technique for persons exposed to these compounds. The use of DNA adduct monitoring for human risk assessment gives better prediction capacity than external or internal dose monitoring, since DNA adduct formation is a step in the process of chemical carcinogenesis. More research, however, is required before DNA adduct measurement can be confidently used to predict human risk from exposure to PACs, although recent research has shown a dose-response relationship between PAHs and cancer.

A number of techniques have been developed for the determination of DNA adducts, with the most sensitive being the ^{32}P-postlabeling assay. Although offering detection limits in the range of one adduct in 10^9 nucleotides, this technique has a limited capacity for adduct identification. If positive identification is required, then techniques such as GC-MS may be used.

The monitoring of protein adducts can be used as a surrogate for DNA adduct monitoring, having a longer life since they are not subject to DNA repair. However, protein adducts are not applicable to mechanistic studies of cancer formation by PACs.

The use of QSARs for prediction of the carcinogenic potency of PACs is well established. Mechanistic approaches based on the formation and reactivity of ultimate carcinogens have traditionally been used, but careful consideration needs to be given to the various stages involved, including transport of parent and activated compounds to sites of action plus all the metabolic activation steps and any deactivation processes that may inhibit carcinogenicity. No single descriptor even at the level of the ultimate carcinogens has been found to be applicable. In many cases, QSARs may only apply to the particular systems studied.

Further research is still required in order to adequately understand the chemical carcinogenesis of PACs. For instance, the structural features of adducts that activate protooncogenes and deactivate tumor-suppression genes are poorly understood. Further work is required to develop QSARs for many of the nonalternant PAHs and heterocyclic and substituted PACs.

Because of the large number of PACs that may be activated or deactivated by different mechanisms and the individual specificities involved in DNA adduct formation, multiple mechanisms will have to be elucidated and complex QSARs may have to be developed to predict the formation of cancer by PACs.

Summary

Chemical carcinogenesis is a multistage process that includes initiation, promotion, and progression. Some carcinogenic PACs have been shown to activate proto-oncogenes and deactivate tumor-suppression genes in the carcinogenic process. The function of DNA repair processes appears to be changed in some cases by PACs. Many PACs are well known for their carcinogenic activity, but for this activity to be exerted, metabolic activation by microsomal enzymes must occur.

The enzyme system responsible for PAC activation is the mixed-function oxidase system and, in particular, cytochrome P-450. In the case of PAHs, oxidation predominately produces reactive diol-epoxides that can then be converted to carbonium ions as the reactive eletrophiles that can then covalently bind to DNA. Regions of high activity exist in PAHs, namely, the "bay," "K," and "L" regions which are associated with π electron distribution. The diol-epoxides can exist in either *syn* or *anti* forms, each of which has two enantiomers producing four stereoisomers in all. Energy considerations favor the formation of the *anti* form. Nitrogen-containing PACs can be metabolically activated in a manner similar to that for PAHs, or the nitrogen atom can be oxidized to form hydroxylamines. These reactive electrophiles can then form covalently bound DNA adducts.

The monitoring of DNA adducts has been used in risk assessment for human exposure to PACs. This form of biomonitoring has advantages over the monitoring of external exposure or body levels of the chemicals in question. In the case of PACs, binding to DNA is an important step in the multistage carcinogenic process. The estimation of DNA adducts has been used in the monitoring of humans exposed to PAHs in a wide range of industrial situations. Recent research has shown a dose-response relationship between PAH adduct levels and human cancer, thus developing molecular epidemiology as a relevant science for the field of risk assessment.

Techniques have been developed for the determination of DNA adducts and these include immunochemical, fluorescence spectroscopic, GC-MS, and ^{32}P-postlabeling methods. The ^{32}P-postlabeling assay is by far the most sensitive, with limits of detection being of the order of one adduct in 10^{10} normal nucleotides. The use of HPLC for separation of adducted nucleotides in this postlabeling assay is becoming more common and gives better resolution of adducts than does the TLC technique used in the traditional assay.

The detection of adducts on hemoglobin and other proteins has been used as a surrogate for DNA adduct estimation. This technique has the advantages of easy accessability of tissue samples and the relative stability of protein adducts. The main disadvantage of this approach, however, is that it is not a measure of genotoxicity. Techniques for the determination of protein adducts include immunoassay, fluorescence spectroscopy, and GC-MS.

The use of QSARs for estimating the toxicity of PACs has been established for a number of years, with the most notable being the bay-region theory of PAH carcinogenesis. This approach uses the fact that bay-region diol-epoxides produce higher π orbital delocalization energies and thus higher stability for the carbocation produced from ring opening. Since then, many QSARs have been established based on molecular orbital calculation of diol-epoxide reactivity or carbocation stability. Physicochemical properties have also been included as descriptors in QSARs, especially those related to water solubility. It appears that the conformation of DNA adducts affects their ability to cause tumors, and QSARs to predict the carcinogenicity of PACs could well include descriptors related to reactivity, distribution, and conformation.

References

Agarwal SK, Sayer JM, Yeh HJC, Pannell LK, Hilton BD, Pigott MA, Dipple A, Yagi H, Jerina DM (1987) Chemical characterization of DNA adducts derived from the configurationally isomeric benzo[c]phenanthrene-3,4-diol 1,2-epoxides. J Am Chem Soc 109:2497-2504.

Amin S, Camanzo J, Hecht SS (1982) Identification of metabolites of 5,11-dimethylchrysene and 5,12-dimethylchrysene and the influence of a peri-methyl group on their formation. Carcinogenesis 3:1159-1162.

Andersen ME, Kushnan K, Conolly RB, McClellan RO (1992) Mechanistic toxicology research and biologically based modelling:Partners for improving quantitative risk assessments. CIIT Activities 12:1-7.

Anderson MW, Reynolds SH, You M, Maronpot RM (1992) Role of proto-oncogene activation in carcinogenesis. Environ Hlth Perspect 98:13-24.

Au WW (1991) Monitoring human populations for effects of radiation and chemical exposures using cytogenetic techniques. Occupational Medicine: State of the Art Previews 6:597-611.

Austin AC, Claxton LD, Lewtas J (1985) Mutagenicity of the fractionated organic emissions for diesel, cigarette smoke condensate, coke oven and roofing tar in the Ames assay. Environ Mut 7:471-487.

Baird WM, Dipple A, Grover PL, Sims P, Brookes P (1973) Studies of the formation of hydrocarbon-deoxyribonucleoside products by the binding of derivatives of 7-methylbenz[a]anthracene to DNA in aqueous solution and in mouse embroyo cells in culture. Cancer Res 33:2386-2392.

Baird WM, Harvey RG, Brookes P (1975) Comparison of the cellular DNA-bound products of benzo[a]pyrene with the products formed by reaction of benzo[a]pyrene-4,5-oxide with DNA. Cancer Res 35:54-47.

Barrett JC (1992) Mechanisms of action of known human carcinogens. In: Vainio H, Magee PN, McGregor DB, McMichael AJ (eds), Mechanisms of carcinogenesis in risk assessment. Internat Agency for Res on Cancer, Lyon, France, pp 115-134.

Beach AC, Gupta RC (1992) Human biomonitoring and the ^{32}P-postlabeling assay. Carcinogenesis 13:1053-1074.

Beland FA, Kadlubar FF (1990) Metabolic activation and DNA adducts of aromatic amines and nitroaromatic hydrocarbons. In: Cooper CS, Grover PC (eds), Chemical carcinogenesis and mutagenesis I. Handbook of experimental pharmacology, vol 94/I. Springer-Verlag, London, pp 267-325.

Belinsky SA, White CM, Devereaux TR, Anderson MW (1987) DNA adducts as a dosemeter for risk estimation. Environ Hlth Perspect 76:3-8.

Benigni R, Guiliani A (1987) Carcinogenicity, mutagenicity, toxicity and chemical structure in a homogenous data base. In: Hadzi D, Jerman-Blazic B (eds), QSAR in drug design and toxicology. Elsevier, Amsterdam, pp 346-348.

Benigni R, Andreoli C, Guiliani A (1989) Structure-activity studies of chemical carcinogens: Use of an electrophilic reactivity parameter in a new QSAR model. Carcinogenesis 10:55-61.

Bjorseth A, Becher G (1986) PAH in work atmospheres: Occurrence and determination. CRC Press, Boca Raton, FL, pp 103-115.

Bolt HM, Peter H, Foest V (1988) Analysis of macromolecular ethylene oxide adducts. Int Arch Occup Environ Hlth 60:141-144.

Bos JL, van Kreyl CF (1992) Genes and gene products that regulate proliferation and differentiation: Critical targets in carcinogenesis. In: Vainio H, Magee PN, McGregor DB, McMichael AJ (eds), Mechanisms of carcinogenesis in risk assessment. Internat Agency for Res on Cancer, Lyon, France, pp 57-65.

Bowden GT, Krieg P (1991) Differential gene exposure during multistage carcinogenesis. Environ Hlth Perspect 93:51-56.

Braun AG, Wornat MJ, Mitra A, Sarofin AF (1987) Environ Hlth Perspect 73: 215-221.

Butlin HT (1892) Cancer of the scrotum in chimney sweeps and others II. Why former sweeps do not suffer from scrotal cancer. Brit Med J 2:1-3.

Chadha A, Sayer JM, Yeh HJC, Yagi H, Cheh AM, Pannell LK, Jerina DM (1989) Structures of covalent nucleoside adducts formed from adenine, guanine and cytosine bases of DNA and the optically active bay-region 3,4-diol 1,2-epoxides of dibenz[a,j] anthracene. J Am Chem Soc 111:5456-5463.

Chang RL, Levin W, Wood AW, Kumar S, Yagi H, Jerina DM, Lehr RE, Conney AH (1984) Tumorigenicity of dihydrodiols and diol-epoxides of benz(c)acridine in new-born mice. Cancer Res 44:5161-5164.

Cody V (1985) Conformational analysis of environmental agents: Use of X-ray crystallographic data to determine molecular reactivity. Environ Hlth Perspect 61:163-183.

Coles B (1984) Effects of modifying structure on electrophilic reactions with biological nucleophiles. Drug Metab Rev 15:1307-1309.

Cook JW, Hewett CL, Hieger I (1933) The isolation of a cancer producing hydrocarbon from coal tar. Parts I, II and III. J Chem Soc: 395-405.

Cushnir JR, Lamb JH, Parry A, Farmer PB (1991) Tandem mass spectrometric approaches for determining exposure to alkylating agents. In: O'Neill IK, Chen J, Bartsch H (eds), Relevance to human cancer of N-nitroso compounds, tobacco smoke and mycotoxins. Internat Agency for Res on Cancer, Lyon, France, pp 107-112.

Daudel P, Duquense M, Vigny P, Grover PL, Sims P (1975) Fluorescene spectral evidence that benzo[a]pyrene-DNA products in mouse skin arise from diolepoxides. FEBS Lett 57:250-253.

Debnath AK, Lopez de Compadre RL, Debnath G, Shusterman AJ, Hansch C (1991) Structure-activity relationships of mutagenic aromatic and heteroaromatic nitro compounds. Correlation with molecular orbital energies and hydrophobicity. J Med Chem 34:786-797.

Delcos KB, El Bayoumy K, Hecht SS, Walker RP, Kadlubar FF (1988) Metabolism of the carcinogen [^{3}H]-8-nitrochysene in the preweanling mouse: Identification of 6-aminochrysene-1,2-dihydrodiol as the probable proximate carcinogenic metabolite. Carcinogenesis 9:1875-1884.

Dewar MJS (1969) The molecular orbital theory of organic chemistry. McGraw Hill, New York, pp 153-189.

Dewar MJS, Dougherty RC (1975) The PMO theory of organic chemistry. Plenum Press, New York, pp 78-85.
Dino JJ, Guenat CR, Tomer KB, Kaufman KB (1987) Analysis of carcinogen modified oligonucleotides by fast atom bombardment/tandem mass spectrometry. Rapid Commun Mass Spec 1:69-71.
Dipple A, Lawley PD, Brookes P (1968) Theory of tumour initiation by chemical carcinogens: Dependence of activity on structure of ultimate carcinogen. Eur J Cancer 4:493-506.
Dipple AD, Moschel RC, Bigger CAH (1984) Polynuclear aromatic carcinogens. In: Searle CE (ed), Chemical carcinogens. Mono 182, American Chemical Society, Washington, DC, pp. 493-506.
Dyke GW, Craven JJ, Hall R, Garner RC (1992) Smoking-related DNA adducts in human gastric cancers. Int J Cancer 52:847-850.
ECETOC (1986) Structure-activity relationships in toxicology and ecotoxicology: An assessment. Mono no. 8., Brussels, pp 34-37.
ECETOC (1989) DNA and protein adducts: Evaluation of their use in exposure monitoring and risk assessment. Mono no. 13, Brussels, p 1089.
Evans HJ, Prosser J (1992) Tumor-suppressor genes: Cardinal factors in inherited predisposition to human cancers. Environ Hlth Perspect 98:25-37.
Farmer PB, Neumann MG, Menschler D (1987) Estimation of exposure in man to substances reacting covalently with macromolecules. Arch Toxicol 60:251-260.
Farmer PB, Lamb J, Lawley PD (1988) Novel uses of mass spectrometry in studies of adducts of alkylating agents with nucleic acids and proteins. In: Bartsch H, Hemminki K, O'Neill IK (eds), Methods for detecting DNA damaging agents in humans: applications in cancer epidemiology and prevention. IARC Scientific Publ 89:347-355.
Farmer PB, Bailey E, Naylor S, Anderson D, Brooks A, Cushnir J, Lamb JH, Sepai O, Tang YS (1993) Identification of endogenous electrophiles by means of mass spectrometric determination of protein and DNA adducts. Environ Hlth Perspect 99:19-24.
Filov VA, Ivin BA (1985) QSAR: Carcinogenic effect of xenobiotics. In: Tichy M (ed), QSAR in toxicology and xenobiochemistry. Elsevier, Amsterdam, pp 99-109.
Friedberg T, Siegert P, Grassow MA, Bartlomowicz B, Oesch F (1990) Studies of the expression of the cytochrome P450IA, P450IIB and P450IIC gene family in extrahepatic and hepatic tissues. Environ Hlth Perspect 88:67-70.
Gaylor DW, Kadlubar FF, Beland FA (1992) Application of biomarkers to risk assessment. Environ Hlth Perspect 98:139-141.
Geactinov NE (1988) Mechanisms of reaction of polycyclic aromatic epoxide derivatives with nucleic acids. In: Yang SK, Silverman BD (eds), Polycyclic aromatic hydrocarbon carcinogenesis: Structure-activity relationships, vol 1. CRC Press, Boca Raton, FL, pp 182-206.
Glatt HR, Mertes I, Wolfel J, Oesch F (1984) Epoxide hydrolases in laboratory animals and in man. In: Greim H, Jung R, Kramer M, Marquardt H, Oesch F (eds), Biochemical basis of chemical carcinogenesis. Raven Press, New York, pp 107-121.
Glauert HP, Schwarz M, Pitot HC (1986) The phenotypic stability of altered hepatic foci: Effect of the short-term withdrawal of phenobarbitol and of the long-term

carcinogens: Use of an electrophilic reactivity parameter in a new QSAR model. Carcinogenesis 10:55-61.

Bjorseth A, Becher G (1986) PAH in work atmospheres: Occurrence and determination. CRC Press, Boca Raton, FL, pp 103-115.

Bolt HM, Peter H, Foest V (1988) Analysis of macromolecular ethylene oxide adducts. Int Arch Occup Environ Hlth 60:141-144.

Bos JL, van Kreyl CF (1992) Genes and gene products that regulate proliferation and differentation: Critical targets in carcinogenesis. In: Vainio H, Magee PN, McGregor DB, McMichael AJ (eds), Mechanisms of carcinogenesis in risk assessment. Internat Agency for Res on Cancer, Lyon, France, pp 57-65.

Bowden GT, Krieg P (1991) Differential gene exposure during multistage carcinogenesis. Environ Hlth Perspect 93:51-56.

Braun AG, Wornat MJ, Mitra A, Sarofin AF (1987) Environ Hlth Perspect 73: 215-221.

Butlin HT (1892) Cancer of the scrotum in chimney sweeps and others II. Why former sweeps do not suffer from scrotal cancer. Brit Med J 2:1-3.

Chadha A, Sayer JM, Yeh HJC, Yagi H, Cheh AM, Pannell LK, Jerina DM (1989) Structures of covalent nucleoside adducts formed from adenine, guanine and cytosine bases of DNA and the optically active bay-region 3,4-diol 1,2-epoxides of dibenz[*a,j*] anthracene. J Am Chem Soc 111:5456-5463.

Chang RL, Levin W, Wood AW, Kumar S, Yagi H, Jerina DM, Lehr RE, Conney AH (1984) Tumorigenicity of dihydrodiols and diol-epoxides of benz(*c*)acridine in new-born mice. Cancer Res 44:5161-5164.

Cody V (1985) Conformational analysis of environmental agents: Use of X-ray crystallographic data to determine molecular reactivity. Environ Hlth Perspect 61:163-183.

Coles B (1984) Effects of modifying structure on electrophilic reactions with biological nucleophiles. Drug Metab Rev 15:1307-1309.

Cook JW, Hewett CL, Hieger I (1933) The isolation of a cancer producing hydrocarbon from coal tar. Parts I, II and III. J Chem Soc: 395-405.

Cushnir JR, Lamb JH, Parry A, Farmer PB (1991) Tandem mass spectrometric approaches for determining exposure to alkylating agents. In: O'Neill IK, Chen J, Bartsch H (eds), Relevance to human cancer of N-nitroso compounds, tobacco smoke and mycotoxins. Internat Agency for Res on Cancer, Lyon, France, pp 107-112.

Daudel P, Duquense M, Vigny P, Grover PL, Sims P (1975) Fluorescene spectral evidence that benzo[*a*]pyrene-DNA products in mouse skin arise from diolepoxides. FEBS Lett 57:250-253.

Debnath AK, Lopez de Compadre RL, Debnath G, Shusterman AJ, Hansch C (1991) Structure-activity relationships of mutagenic aromatic and heteroaromatic nitro compounds. Correlation with molecular orbital energies and hydrophobicity. J Med Chem 34:786-797.

Delcos KB, El Bayoumy K, Hecht SS, Walker RP, Kadlubar FF (1988) Metabolism of the carcinogen [^{3}H]-8-nitrochysene in the preweanling mouse: Identification of 6-aminochrysene-1,2-dihydrodiol as the probable proximate carcinogenic metabolite. Carcinogenesis 9:1875-1884.

Dewar MJS (1969) The molecular orbital theory of organic chemistry. McGraw Hill, New York, pp 153-189.

Dewar MJS, Dougherty RC (1975) The PMO theory of organic chemistry. Plenum Press, New York, pp 78-85.

Dino JJ, Guenat CR, Tomer KB, Kaufman KB (1987) Analysis of carcinogen modified oligonucleotides by fast atom bombardment/tandem mass spectrometry. Rapid Commun Mass Spec 1:69-71.

Dipple A, Lawley PD, Brookes P (1968) Theory of tumour initiation by chemical carcinogens: Dependence of activity on structure of ultimate carcinogen. Eur J Cancer 4:493-506.

Dipple AD, Moschel RC, Bigger CAH (1984) Polynuclear aromatic carcinogens. In: Searle CE (ed), Chemical carcinogens. Mono 182, American Chemical Society, Washington, DC, pp. 493-506.

Dyke GW, Craven JJ, Hall R, Garner RC (1992) Smoking-related DNA adducts in human gastric cancers. Int J Cancer 52:847-850.

ECETOC (1986) Structure-activity relationships in toxicology and ecotoxicology: An assessment. Mono no. 8., Brussels, pp 34-37.

ECETOC (1989) DNA and protein adducts: Evaluation of their use in exposure monitoring and risk assessment. Mono no. 13, Brussels, p 1089.

Evans HJ, Prosser J (1992) Tumor-suppressor genes: Cardinal factors in inherited predisposition to human cancers. Environ Hlth Perspect 98:25-37.

Farmer PB, Neumann MG, Menschler D (1987) Estimation of exposure in man to substances reacting covalently with macromolecules. Arch Toxicol 60:251-260.

Farmer PB, Lamb J, Lawley PD (1988) Novel uses of mass spectrometry in studies of adducts of alkylating agents with nucleic acids and proteins. In: Bartsch H, Hemminki K, O'Neill IK (eds), Methods for detecting DNA damaging agents in humans: applications in cancer epidemiology and prevention. IARC Scientific Publ 89:347-355.

Farmer PB, Bailey E, Naylor S, Anderson D, Brooks A, Cushnir J, Lamb JH, Sepai O, Tang YS (1993) Identification of endogenous electrophiles by means of mass spectrometric determination of protein and DNA adducts. Environ Hlth Perspect 99:19-24.

Filov VA, Ivin BA (1985) QSAR: Carcinogenic effect of xenobiotics. In: Tichy M (ed), QSAR in toxicology and xenobiochemistry. Elsevier, Amsterdam, pp 99-109.

Friedberg T, Siegert P, Grassow MA, Bartlomowicz B, Oesch F (1990) Studies of the expression of the cytochrome P450IA, P450IIB and P450IIC gene family in extrahepatic and hepatic tissues. Environ Hlth Perspect 88:67-70.

Gaylor DW, Kadlubar FF, Beland FA (1992) Application of biomarkers to risk assessment. Environ Hlth Perspect 98:139-141.

Geactinov NE (1988) Mechanisms of reaction of polycyclic aromatic epoxide derivatives with nucleic acids. In: Yang SK, Silverman BD (eds), Polycyclic aromatic hydrocarbon carcinogenesis: Structure-activity relationships, vol 1. CRC Press, Boca Raton, FL, pp 182-206.

Glatt HR, Mertes I, Wolfel J, Oesch F (1984) Epoxide hydrolases in laboratory animals and in man. In: Greim H, Jung R, Kramer M, Marquardt H, Oesch F (eds), Biochemical basis of chemical carcinogenesis. Raven Press, New York, pp 107-121.

Glauert HP, Schwarz M, Pitot HC (1986) The phenotypic stability of altered hepatic foci: Effect of the short-term withdrawal of phenobarbitol and of the long-term feeding of purified diets after the withdrawal of phenobarbitol. Carcinogenesis 7:117-121.

Gold A, Sangaiah R, Nesnow S (1988) Structure-activity relationships in the metabolism and biological activity of cyclopenta-fused polycyclic aromatic systems. In: Yang SK, Silvermann BD (eds), Polycyclic aromatic hydrocarbon carcinogenesis: Structure activity relationships, vol 1. CRC Press, Boca Raton, FL, pp 177-207.

Gorelick NJ, Reeder NL (1993) Detection of multiple polycyclic aromatic hydrocarbon DNA adducts by a high performance liquid chromatography P^{32}-postlabelling method. Environ Hlth Perspect 99:207-212.

Grimmer G, Brune H, Dettbarn G, Jacob J, Misfeld J, Mohr U, Nanjack K, Timm J, Wenzel-Hartung R (1991) Relevance of polycyclic aromatic hydrocarbons as environmental carcinogens. Fresnius J Anal Chem 339:792-795.

Guenther TM, Oesch F (1981) Microsomal epoxide hydrolase and its role in polyclic aromatic hydrocarbon transformation. In: Gelboin HV, Ts'O POP (eds), Polycyclic hydrocarbons and cancer, vol 3. Academic Press, New York, pp 3-42.

Gupta RC, Randerath K (1988) Analysis of DNA adducts by ^{32}P labelling and thin layer chromatography. In: Friedberg EC, Hanawalt PC (eds), DNA repair, vol 3. Marcel Dekker, New York, pp 399-418.

Gupta RC (1988) ^{32}P-adduct assay: Short- and long-term persistence of 2-acetylaminofluorene-DNA adducts and other applications of the assay. Cell Biol Toxicol 4:467-475.

Hall M, Grover PL (1990) Polycyclic aromatic hydrocarbons, metabolism, activation and tumour initiation. In: Cooper CS, Grover PL (eds), Chemical carcinogenesis and mutagenesis I. Handbook of experimental pharmacology, vol 94/1. Springer-Verlag, London, pp 327-372.

Harris CC (1985) Future directions in the use of DNA adducts as internal dosimeters for monitoring human exposure to environmental mutagens and carcinogens. Environ Hlth Perspect 62:185-191.

Harris CC, Vahakangas K, Newmann MJ, Trivers GE, Shamsuddin A, Limopoli N, Mann DL, Wright WE (1985) Detection of benzo[a]pyrene diol epoxide-DNA adducts in peripheral blood lymphocytes and antibodies to the adducts in serum from coke-oven workers. Proc Natl Acad Sci USA 82:6672-6676.

Harris CC (1992) Tumor suppression genes, multistage carcinogenesis and molecular epidemiology. In: Vainio H, Magee PN, McGregor DB, McMichael AJ (eds), Mechanisms of carcinogenesis in risk identification. Internat Agency for Res on Cancer, Lyon, France, pp 67-86.

Harvey RG (1981) Activated metabolites of carcinogenic hydrocarbons. Acc Chem Res 14:218-226.

Haugen DA, Peak MJ, Reilly CA (1981) Chemical and biological characterisation of high-BTU coal gasification. (The HYGAS process). II. Nitrous acid treatment for detection of mutagenic primary aromatic amines: Non-specific reactions. In: Mahlum DD, Gray RH, Felix WD (eds), Coal conversion and the environment. Tech Inform Ctr, US Dept of Energy, Washington, DC, pp 115-127.

Haugen DA, Peak MJ (1983) Mixtures of polycyclic aromatic compounds inhibit mutagenesis in the *Salmonella* microsome assay by inhibition of metabolic activation. Mutat Res 116:257-260.

Haugen D, Becher G, Benestead C, Vahakangas K, Trivers GE, Newman MJ, Harris CC (1986) Determination of polycyclic aromatic hydrocarbons in the urine, benzo[a]pyrene diol epoxide-DNA adducts in lymphocyte DNA and antibodies to the adducts in sera from coke-oven workers exposed to measured amounts of polycyclic aromatic hydrocarbons in the work atmosphere. Cancer Res 46:4178-4183.

Haugen D, Becher G, Benstead C, Vahakangas K, Trivers GE, Newman MJ, Harris CC (1988) Biomonitoring of individuals exposed to high levels of PAH in the work environment. In: Cooke M, Dennis AJ (eds), Polynuclear aromatic hydrocarbons: A decade of progress. Battelle Press, Columbus, OH, pp 377-390.

Haugen DA, Myers SR (1990) Reaction of benzo[a]pyrene-7,8-dihydrodiol-9,10-epoxide with human haemoglobin and chromatographic resolution of the covalent adducts. DE 90 008705, Argonne Nat Lab, Argonne, IL.

Hecht SS, Melikian AA, Amin S (1988) Effects of methyl substitution on the tumorigenicity and metabolic activation of polycyclic aromatic hydrocarbons. In: Yang SK, Silverman BD (eds), Polycyclic aromatic hydrocarbon carcinogenesis: Structure-activity relationships, vol 1. CRC Press, Boca Raton, FL, pp 95-129.

Hemminki K, Grzybowska E, Chorazy M, Twardowska-Saucha K, Sroczynski JW, Putman KL, Randerath K, Phillips DH, Hewer A, Santella RM, Perera FP (1990) DNA adducts in humans related to occupational exposure to aromatic compounds. In: Vainio H, Sorsa M, McMichael AJ (eds), Complex mixtures and cancer risk. IARC scientific publ, no 104, Internat Agency for Res on Cancer, Lyon, France, pp 181-192.

Hemminki K, Reunanen A, Kahn H (1991) Use of DNA adducts in the assessment of occupational and environmental exposure to carcinogens. Eur J Cancer 27:289-291.

Hemminki K (1992) Significance of DNA and protein adducts. In: Vainio H, Magee PN, McGregor DB, McMichael AJ (eds), Mechanisms of carcinogenesis in risk identification. Internat Agency for Res on Cancer, Lyon, France, pp 525-534.

Hendrick S, Glauert HP, Pitot HC (1986) The phenotypic stability of altered hepatic foci: Effects of withdrawal and subsequent readministration of phenobarbitol. Carcinogenesis 7:2041-2045.

Herbert R, Marcus A, Wolff MS, Perera FP, Andrews L, Godbold JH, Rivera M, Stefanidis M, Lu XQ, Landrigan PJ, Santella RM (1990). A pilot study of detection of DNA adducts in white blood cells of roofers by 32P-postlabelling. In: Vainio H, Sorsa M, McMichael AJ (eds), Complex mixtures and cancer risk. IARC scientific publ no 104, Internat Agency for Res on Cancer, Lyon, France, pp 205-214.

Herndon WC (1981) Model calculations for reactivities of polycyclic aromatic hydrocarbon metabolites. Tetrahedron Lett 22:983-986.

Herreno-Saenz D, Evans FE, Lai EC, Abian J, Fu PP, Delclos KB (1993) Products formed from the in vitro reaction of metabolites of 3-aminochrysene with calf thymus DNA. Chem-Biol Interactions 86:1-15.

Hodgson RM, Leidel A, Bochnitschek W, Glatt HR, Oesch F, Grover PL (1986) Metabolism of the bay-region diol epoxide of chrysene to a triol epoxide and the enzyme catalysed conjugation of these epoxides with glutathione. Carcinogenesis 7:2095-2098.

Hruszkewycz AM, Canella KA, Peltenen K, Kotrappa L, Dipple A (1992) DNA polymerase action on benzo[*a*]pyrene-DNA adducts. Carcinogenesis 13:2347–2352.

IARC (1984a) Polynuclear aromatic compounds, part 2. Carbon blacks, mineral oils and some nitroarenes. In: IARC monographs on the evaluation of the carcinogenic risk of chemicals to humans, vol 33. Lyon, France, pp 171–222.

IARC (1984b) Polynuclear aromatic compounds, part 3. Industrial exposures in aluminium production, coal gasification, coke production and iron and steel founding. In: IARC monographs on the evaluation of the carcinogenic risk of chemicals to humans, vol 34. Lyon, France, pp 33–190.

IARC (1985) Polynuclear aromatic compounds, part 4. Bitumins, coal tars and derived products, shale oils-soots. In: IARC monographs on the evaluation of the carcinogenic risk of chemicals to humans, vol 35. Lyon, France, pp 39–240.

IARC (1987) IARC monographs on the evaluation of carcinogenic risk to humans, overall evaluations of carcinogenicity: An updating of IARC monographs, vols 1–42, suppl 7. Lyon, France, pp 40–74.

IARC (1990) Complex mixtures and cancer risk, Vainio H, Sorsa M, McMichael AJ (eds). Internat Agency for Res on Cancer, Lyon, France, pp 1–8.

Izzotti A, Rossi GA, Bagnasco M, de Floa S (1991) Benzo(*a*)pyrene diol-epoxide-DNA adducts in alveolar macrophages of smokers. Carcinogenesis 12:1281–1285.

Jakoby WB, Ketterer B, Mannervik B (1984) Glutathione transferases: Nomenclature. Biochem Pharmacol 33:2539–2540.

Jerina DM, Lehr RE, Yagi H, Hernandez O, Dansette PM, Wislocki PG, Wood AW, Chang RL, Levin W, Conney AH (1976) Mutagenicity of benzo[*a*]pyrene derivatives and the description of a quantum mechanical model which predicts the ease of carbonium ion formation from diol epoxides. In: de Serres FJ, Fonts JR, Bend JR Philpot RM (eds), *In vitro* metabolic activation in mutagenesis testing. Elsevier, Amsterdam, pp 159–177.

Jerina DM, Lehr RE (1977) The bay-region theory: A quantum mechanical approach to aromatic hydrocarbon-induced carcinogenicity. In: Ulrich V (ed), Microsomes and drug oxidations. Pergamon Press, New York, pp 709–714.

Jerina DM, Lehr R, Schaifer-Ridder M, Yagi H, Karle JM, Thakker DR, Wood AW, Lu AYH, Ryan D, West S, Levin W, Conney AH (1977) Bay-region epoxides of dihydrodiols. A concept explaining the mutagenic and carcinogenic activity of benzo[*a*]pyrene and benz[*a*]anthracene. In: Hiatt HH, Watson JD, Winsten JA (eds), Origns of human cancer, book B. Cold Spring Harbor Lab, Cold Spring Harbor, NY, pp 639–658.

Jerina DM, Sayer JM, Yagi H, Croisy-Delcey M, Ittah Y, Thakker DR (1982) Highly tumorigenic bay-region diol epoxides from the weak carcinogen benzo[*c*]pheranthene. In: Snyder R, Parke DV, Kocsis J, Jollow CJ, Gibson GG, Witner CM (eds), Biological reactive intermediates, vol 2. Plenum Press, New York, pp 501–509.

Jerina DM, Yagi H, Thakker DR, Sayer JM, van Bladeren PJ, Lehr RE, Whalen DL, Levin W, Chang RC, Wood AW, Conney AH (1984) Identification of the ultimate carcinogenic metabolites of the polycyclic aromatic hydrocarbons: Bay-region (R,S)-diol-(S,R)-epoxides. In: Caldwell J, Paulson GD (eds), Foreign compound metabolism. Taylor & Francis, London, pp 257–260.

Jernstrom B, Martinez M, Meyer DJ, Ketterer B (1985) Glutathione conjugation of the carcinogenic and mutagenic electrophile (±)-7β,8dihydroxy-9α,10α-oxy-7,8,9,10-tetrahydrobenzo[a]pyrene catalysed by purified rat liver glutathione transferases. Carcinogenesis 6:85–89.

Johnson RW, Vo-Dinh T (1989) Fumed silica substrates for enhanced fluorescence spot test analysis of benzo[a]pyrene-DNA adduct products. Anal Chem 61:2766–2769.

Kadlubar FF, Fu PP, Jung H, Shaikh AU, Beland FA (1990) The metabolic N-oxidation of carcinogenic arylamines in relation to nitrogen charge density and oxidation potential. Environ Hlth Perspect 87:233–236.

Kapitulnik J, Wislocki PG, Levin W, Yagi H, Jerina DM, Conney AH (1978) Tumorigenicity studies with diol-epoxides of benzo(a)pyrene which indicate that (±)trans-7β,8α-dihydroxy-9α,10α-epoxy-7,8,9,10-tetrahydrobenzo[a]pyrene is an ultimate carcinogen in newborn mice. Cancer Res 38:354–357.

Kashino S, Zacharias DE, Prout CK, Carrell HL, Glusker JP, Hecht SS, Harvey RS (1984) Structure of 5-methylchrysene, $C_{19}H_{14}$. Acta Cryst C40: 536–540.

Kennaway EL, Hieger I (1930) Carcinogenic substances and their fluorescence spectra. Brit Med J 1:1044–1046.

King LC, Kohan MJ, George SE, Lewtas J, Claxton LD (1990) Metabolism of 1-nitropyrene by human, rat and mouse intestinal flora: Mutagenicity of isolated metabolites by direct analysis of HPLC fractions with a microsuspension reverse mutation assay. J Toxicol Environ Hlth 31:179–192.

Kinzel V, Furstenberger G, Loehrke H, Marks F (1986) Three stage tumorigenesis in mouse skin: DNA synthesis as a prerequisite for the conversion stage induced by TPA prior to initiation. Carcinogenesis 7:779–782.

Klein G (1988) Cellular oncogene activation. Marcel Dekker, New York.

Klopman G (1985) Predicting toxicity through a computer automated structure evaluation program. Environ Hlth Perspect 61:269–274.

Klopman G, Rosenkranz H (1991) Structure-activity relations: Maximizing the usefulness of mutagenicity and carcinogenicity data bases. Environ Hlth Perspect 96:67–75.

Kriek E, Vanschooten FJ, Hillebrand MJX, Vanleeuwen FE, Denengelse L, Delooff AJA, Dijkmans APG (1993) DNA adducts as a measure of lung cancer risk in humans exposed to polycyclic aromatic hydrocarbons. Environ Hlth Perspect 99:71–76.

Lau HS, Baird WM (1991) Detection and identification of benzo[a]pyrene-DNA adducts by [^{35}S] phosphorothioate labelling and HPLC. Carcinogenesis 12:885–893.

Lau HS, Baird WM (1992) Human cell metabolism and DNA adduction of polycyclic aromatic hydrocarbons. In: Milo EC, Castro BC, Shirler CF (eds), Transformation of human epithelial cells, CRC Press, Boca Raton, FL, pp 31–66.

Lecoq S, Pfau W, Grover PL, Phillips DH (1992) HPLC separation of ^{32}P-postlabelled DNA adducts formed from dibenz[a,h]anthracene in skin. Chem-Biol Interactions 85:173–185.

Lee BM, Baoyun MS, Herbert R, Hemminki K, Perera FP, Santella RM (1991) Immunologic measurement of polycyclic aromatic hydrocarbon albumin adducts in foundry workers and roofers. Scand J Work Environ Hlth 17:190–194.

Lehr RE, Kumar S, Levin W, Wood AW, Chang RC, Conney AH, Yagi H, Sayer JM, Jerina DM (1985) The bay-region theory of polycyclic aromatic hydrocarbon carcinogenesis. In: Harvey RG (ed), Polycyclic aromatic hydrocarbons and carcinogenesis, ACS Symp Series 283, American Chemical Society, Washington, DC, pp 63–69.

Lehr RE, Wood AW, Levin W, Conney AH, Jerina DM (1988) Benzacridines and dibenzacridines: Metabolism, mutagenicity and carcinogenicity. In: Yang SR, Silverman BD (eds), Polycyclic aromatic hydrocarbon carcinogenesis: Structure-activity relationships, vol 1. CRC Press, Boca Raton, FL, pp 31–58.

Levin W, Wood AW, Chang RC, Kimar S, Yagi H, Jerina DM, Lehr RE, Conney AH (1983) Tumor initiating activity of benz(c)acridine and twelve of its derivatives on mouse skin. Cancer Res 43:4625–4627.

Loew GH, Sudhindra BS, Ferrell JE (1979) Quantum chemical studies of polycyclic aromatic hydrocarbons and their metabolites: correlations to carcinogenicity. Chem-Biol Interactions 26:75–89.

Loew GH, Poulsen M, Kirhjian E, Ferrell J, Sudhindra S, Rebagliati M (1985) Environ Hlth Perspect 61:69–96.

Lohman PHM (1988) Summary: Adducts. In: Bartsch H, Hemminki K, O'Neill IK (eds), Methods for detecting DNA damaging agents in humans: Applications in cancer epidemiology and Prevention. IARC scientific publ no 89, Internat Agency for Res on Cancer, Lyon, France, pp 13–20.

Lowe JP, Silverman BD (1983) Heteroatom effects in chemical carcinogenesis: Effects of ring heteroatoms on ease of carbocation formation. Cancer Biochem Biophys 7:53–56.

Lutz WK (1979) *In vivo* covalent binding of organic chemicals to DNA as a quantitative indicator in the process of chemical carcinogenesis. Mutat Res 65:289–356.

Maher VM, Patton JD, Yang JL, Wang Y, Yang LL, Aust AE, Bhattachajya N, McCormic JJ (1987) Mutations and homologous recombination induced in mammalian cells by metabolites of benzo[a]pyrene and 1-nitropyrene. Environ Hlth Perspect 76:33–39.

Mallet WG, Mosebrook DR, Trush MA (1991) Activation of ($\pm$) -*trans*-7,8-dihydroxy-7,8-dihyd[o]enzo(a)pyrene to diolepoxides by human polymorphonuclear leukocytes on myeloperoxidase. Carcinogenesis 12:521–524.

Manchester DK, Wilson VL, Hsu IC, Choi JS, Parker NB, Mann DL, Weston A, Harris CC (1990) Synchronous fluorescence spectroscopic, immunoaffinity chromatographic and ^{32}P-postlabelling analysis of human placenta DNA known to contain benzo[a]pyrene diol epoxide adducts. Carcinogenesis 11:553–559.

Mannervik B, Jensson H (1982) Binary combinations of four protein sub-units with different catalytic specificities explain the relationship between six basic glutathione S-transferases in rat liver cytosol. J Biol Chem 257:9909–9912.

Miertus S, Trebaticka M, Frecer V (1985) Studies of the QSAR and mechanisms of the action of mutagenic and carcinogenic compounds based on quantum chemical calculations. In: Tichy M (ed), QSAR in toxicology and xenobiochemistry. Elsevier, Amsterdam, pp 127–141.

Minnetian O, Saha M, Giese RW (1987) Oxidation-elimination of a DNA base from its nucleoside to facilitate determination of alkyl chemical damage to DNA by gas chromatography with electrophore detection. J Chromatogr 410:453–457.

Mitchell CS, Klopman G, Rosenkranz HS (1986) Computer automated evaluation of mutagenicity and carcinogenicity of selected polycyclic aromatic hydrocarbons. In: Cooke M, Dennis AJ (eds), Polynuclear aromatic hydrocarbons:

Chemistry, characterization and carcinogenesis. Battelle Press, Columbus, OH, pp 611–624.

Moller L, Zeisig M (1993) DNA adduct formation after oral administration of 2-nitrofluorene and N-acetyl-2-aminofluorene, analysed by ^{32}P-TLC and ^{32}P-HPLC. Carcinogenesis 14:53–59.

Montesano R (1990) Approaches to detecting individual exposure to carcinogens. In: Vainio H, Sorsa M, McMichael AJ (eds), Complex mixtures and cancer risk. IARC scientific publ no 104, Internat Agency for Res on Cancer, Lyon, France, pp 11–19.

Mullaart E, Bourigter METI, Lohman PHM, Vijg J (1989) Age-related induction and disappearance of carcinogen-DNA-adducts in livers of rats exposed to low levels of 2-acetylaminofluorene. Chem-Biol Interactions 69:373–384.

Mumford JL, Lee XM, Lewtas J, Young TL, Santella RM (1993) DNA adducts as biomarkers for assessing exposure to polycyclic aromatic hydrocarbons in tissues from Xuan-Wei women with high exposure to coal combustion emissions and high lung cancer mortality. Environ Hlth Perspect 99:83–88.

National Academy of Sciences (1983) Risk assessment in federal government: Managing the process. National Academy Press, Washington, DC.

Neumann HG. (1988) Haemoglobin binding in control of exposure to and risk assessment of aromatic amines. In: Bartsch H, Hemminki K, O'Neill IK (eds), Methods for detecting DNA damaging agents in humans: Applications in cancer epidemiology and prevention. IARC scientific publ, no 89, Internat Agency for Res on Cancer, Lyon, France, pp 157–165.

Neumann HG, Hammerl R, Hillesheim W, Wildschutte M (1990) Role of genotoxic and nongenotoxic effects in multistage carcinogenicity of aromatic amines. Environ Hlth Perspect 88:207–212.

Orzechowski A, Schrenk D, Bock KW (1992). Metabolism of 1- and 2-naphthylamine in isolated rat hepatocytes. Carcinogenesis 13:2227–2232.

Osborne NR (1979) Carcinogenicity indices in polycyclic hydrocarbons. Cancer Res 39:4760–4761.

Ovrebo S, Hewer A, Phillips DH, Haugen A (1990) Polycyclic aromatic hydrocarbon-DNA adducts in coke-oven workers. In: Vainio H, Sorsa M, McMichael AJ (eds), Complex mixtures and cancer risk. IARC scientific publ no 104, Internat Agency for Res on Cancer, Lyon, France, pp 193–198.

Patton JD, Maher VM, McCormick JJ (1986) Cytotoxicity, mutagenicity and transformation of duploid human fibroblasts by 1-nitropyrene and 1-nitrosopyrene. In: Cooke M, Dennis AJ (eds), Polynuclear aromatic hydrocarbons: A decade of progress. Battelle Press, Columbus, OH, pp 687–697.

Paulius DE, Prakash AS, Harvey RG, Ambrovich M, LeBreton PR (1986) 1-Alkyl substitution effects on the DNA intercalation of benzo(a)pyrene metabolites. In: Cooke M, Dennis AJ (eds), Polynuclear aromatic hydrocarbons: Chemistry, characterization and carcinogenesis, Battelle Press, Columbus, OH, pp 745–754.

Perera F, Jeffrey A, Santella RM, Brenner D, Mayer J, Latriano L, Smith S, Young TL, Tsai WY, Hemminki K, Brandt-Rauf P (1990) Macromolecular adducts and related biomarkers in biomonitoring and epidemiology of complex mixtures. In: Vainio H, Sorsa M, McMichael AJ (eds), Complex mixtures and cancer risk. IARC scientific publ no 104, Internat Agency for Res on Cancer, Lyon, France, pp 164–180.

Perera F, Brenner D, Jeffrey A, Mayer J, Tang D, Warburton D, Young TI, Waznek L, Latriono L, Motykiewicz G, Grzybowska E, Chorazy M, Hemminki K, Santella R (1992) DNA adducts and related biomarkers in populations exposed to environmental carcinogens. Environ Hlth Perspect 98:133-137.

Perera FP, Santella RM, Brenner D, Poirier MC, Munshi AA, Fischman HK, van Ryzin J (1987) DNA adducts, protein adducts and sister chromatid exchange in cigarette smokers and non-smokers. J Natl Cancer Inst 79:449-456.

Perera FP, Hemminki K, Young TL, Brenner D, Kelly G, Santella RM (1988) Detection of polycyclic aromatic hydrocarbon-DNA adducts in the white blood cells of foundry workers. Cancer Res 48:2288-2291.

Phillips DH (1990) Modern methods of DNA adduct determination. In: Cooper CS, Grover PL (eds), Chemical carcinogenesis and mutagenesis I. Handbook of experimental pharmacology, vol 94/I. Springer-Verlag, London, pp 503-546.

Phillips DH, Schoket B, Hewer A, Grover P (1990) DNA adduct formation in human and mouse skin by mixtures of polycyclic aromatic hydrocarbons. In: Vainio H, Sorsa M, McMichael AJ (eds), Complex mixtures and cancer risk. IARC scientific publ no 104, Internat Agency for Res on Cancer, Lyon, France, pp 223-232.

Phillips DH (1992) Use of ^{32}P-postlabelling to distinguish between genotoxic and non-genotoxic carcinogens. In: Vainio H, Magee PN, McGregor DB, McMichael AJ (eds), Mechanisms of carcinogenesis in risk identification. Internat Agency for Res on Cancer, Lyon, France, pp 211-224.

Phillips DH, Hewer A (1993) DNA adducts in human urinary bladder and other tissues. Environ Hlth Perspect 99:45-50.

Pitot HC (1986) Fundamentals of oncology, 3rd ed. Marcel Dekker, New York.

Pitot HC (1990) Mechanisms of chemical carcinogenesis: Theoretical and experimental bases. In: Cooper CS, Growler PL (eds), Chemical carcinogenesis and mutagenesis I. Handbook of experimental pharmacology, vol 94/I. Springer-Verlag, London, pp 2-29.

Platt KL, Schollmeier M, Frank H, Oesch F (1990) Stereoselective metabolism of dibenz[a]h)anthracene to trans-dihydrodiols and their activation to bacterial mutagens. Environ Hlth Perspect 88:37-42.

Pott P (1775) Chirurgical observations relative to the cataract, the polypus of the nose, the cancer of the scrotum, the different kinds of ruptures and the mortification of the toes and feet. Hawes, Clarke & Collins, London.

Pott F, Heinrich U (1990) Relative significance of different hydrocarbons for the carcinogenic potency of emissions from various incomplete combustion processes. In: Vainio H, Sorsa M, McMichael AJ (eds), Complex mixtures and cancer risk. IARC Scientific Publ 106:288-297.

Prakash AS, Harvey RG, LeBreton PR (1988) Differences in the influence of π physical binding interactions with DNA on the reactivity of bay versus K-region hydrocarbon epoxides. In: Cooke M, Dennis AJ (eds), Polynuclear aromatic hydrocarbons: A decade of progress. Battelle Press, Columbus, OH, pp 699-710.

Pullman A, Pullman B (1955) Electronic structure and carcinogenic activity of aromatic molecules, new developments. Adv Cancer Res 3:117-169.

Rall DP (1990) Carcinogens in our environment. In: Vainio H, Sorsa M, McMichael AJ (eds), Complex mixtures and cancer risk. IARC scientific publ no 104, Internat Agency for Res on Cancer, Lyon, France, pp 233-239.

Randerath K, Randerath E, Agrarwal HP, Gupta RC, Schurdak ME, Reddy MV (1985) Postlabelling methods for carcinogen−DNA adduct analysis. Environ Hlth Perspect 62:57-65.

Randerath K, Liehr A, Gladek A, Randerath E (1989) Age-dependent covalent DNA alterations (I compounds) in rodent tissues: Species, tissue and sex specificities. Mutat Res 219:121-133.

Randerath K, Li D, Randerath E (1990) Age-related DNA modfications (I compounds): Modulation by physiological and pathological processes. Mutat Res 238:245-254.

Randerath K, Randerath E (1991) ^{32}P-postlabelling analysis of mutagen/carcinogen-DNA adducts. Pure Appl Chem 63:1283-1286.

Randerath K, Putman KL, Osterburg HH, Johnson SA, Morgan DG, Finch CE (1993) Age-dependent increases of DNA adducts (I compounds) in human and rat brain DNA. Mutat Res 295:11-18.

Reddy MV, Randerath K (1987) ^{32}P-postlabelling assay for carcinogen-DNA adducts: Nuclease P_1-mediated enhancement of its sensitivity and applications. Environ Hlth Perspect 76:41-47.

Reddy AC, Caldwell M, Filakow PJ (1987) Studies of skin tumorigenesis in PGK mosaic mice: Many promoter-independent papillomas and carcinomas do not develop from the existing promoter-dependent papillomas. Int J Cancer 29:261-265.

Risse G, Neuberg M, Hunter JB, Verrier B, Muller R (1990) Products of the *fos* and *jun* proto-oncogenes bind cooperatively to the API DNA recognition sequence. Environ Hlth Perspect 88:133-140.

Rodriguez H, Loechler EL (1993) Mutagenesis by the ($\pm$)-*anti*-diol epoxide of benzo[*a*]pyrene: What controls mutagenic specificity. Biochemistry 32:1759-1769.

Rojas M, Alexandrov K (1986) *In vivo* formation and persistence of DNA adducts in mouse and rat skin exposed to ($\pm$)*trans*-7,8-dihydroxy-7,8-dihydrobenzo[*a*]pyrene and ($\pm$)-7β, 8α-dihydroxy-9α,10α-epoxy-7,8,9,10-tetrahydrobenzo[*a*]pyrene. Carcinogenesis 7:1553-1560.

Rosenkranz HS, Mermelstein R (1985) The mutagenic and carcinogenic properties of nitrated polycyclic aromatic hydrocarbons. In: White CM (ed), Nitrated polycyclic aromatic hydrocarbons. Dr Alfred Huething Verlag, Heidelberg, pp 267-297.

Rosenkranz HS (1992) Structure-activity relationships for carcinogens with different modes of action. In: Vainio H, Magee PN, McGregor DB, McMichael AJ (eds), Mechanisms of carcinogenesis in risk identification. Internat Agency for Res on Cancer, Lyon, France, pp 271-277.

Roussel OP, Chalvert O, Ekert B, Lhoste JM, Mispelter J, Sanguem S, Zajdela F (1988) Dibenzofluoranthenes, their biological activities, structure activity relationships and metabolic activation. In: Yang SK, Silvermann BD (eds), Polycyclic aromatic hydrocarbon carcinogenesis: Structure-activity relationships, vol 1 CRC Press, Boca Raton, FL, pp 67-88.

Roy AK, El-Bayoumy K, Hecht SS (1989) ^{32}P-postlabelling analysis of 1-nitropyrene-DNA adducts in female Sprague-Dawley rats. Carcinogenesis 10: 195-198.

Roy AK, Upadhyaya P, Evans FE, El Bayoumy K (1991a) Structural characteriza-

tion of the major adducts formed by reaction of 4,5-epoxy-4,5-dihydro-1-nitropyrene with DNA. Carcinogenesis 12:577-581.
Roy AK, Upadhyaya P, Fu PP, El-Bayoumy K (1991b) Identification of the major metabolites and DNA adducts formed from 2-nitropyrene *in vitro*. Carcinogenesis 12:475-479.
Royer RE, Mitchell CE, Hanson RL, Dutcher IS, Bechtold WE (1983) Fractionation, chemical analysis and mutagenicity testing of low-BTU coal gasifer tar. Environ Res 31:460-471.
Sanders MJ, Cooper RS, Jankoviak R, Small GJ, Heisig V, Jeffrey AM (1986) Identification of polycyclic aromatic hydrocarbon metabolites and DNA adducts in mixtures using fluorescence line narrowing spectrometry. Anal Chem 58:816-820.
Santella RM, Lin CD, Cleveland WL, Weinstein IB (1984) Monoclonal antibodies to DNA modified by a benzo[*a*]pyrene diol epoxide. Carcinogenesis 5:373-377.
Santella RM, Hsieh LL, Lin CD, Viet S, Weinstein IB (1985) Quantitation of exposure to benzo[*a*]pyrene with monoclonal antibodies. Environ Hlth Perspect 62:95-99.
Santella RM, Lin CD, Dharmaraja M (1986) Monoclonal antibodies to a benzo[*a*]pyrene diol epoxide-modified DNA. Carcinogenesis 7:441-444.
Santella RM, Weston A, Perera FP, Trivers GE, Harris CC, Young TL, Nguyen D, Lee BM, Poirier MC (1988) Inter-laboratory comparison of anti-sera and immunoassays for benzo[*a*]pyrene diol epoxide modified DNA. Carcinogenesis 9:1265-1269.
Santella RM (1991) Adducts as biomarkers of exposure to environmental and occupational carcinogens. Environ Carcino and Ecotox Res C9(1):57-81.
Santella RM, Hemminki K, Tang DL, Paik M, Ottman R, Young TL, Savela K, Vodichova L, Dickey C, Whyatt R, Perera F (1993) Polycyclic aromatic hydrocarbon-DNA adducts in white blood cells and urinary 1-hydroxypyrene in foundry workers. Cancer Epidemiology Biomarkers and Prevention 2:59-62.
Scharping CE, McManus ME, Holder GM (1992) NADPH-supported and arachidonic acid-supported metabolism of the enantiomers of *trans*-7,8-dihydrobenzo[*a*]pyrene-7,8-diol by human liver microsomal samples. Carcinogenesis 13:1199-1207.
Schulte-Hermann R (1985) Tumor promotion in the liver. Arch Toxicol 57:147-158.
Scicchitano DA, Hanawalt PC (1992) Intragenomic repair heterogenicity of DNA damage. Environ Hlth Perspect 98:45-51.
Seybold PG (1986) Steric and electronic determinants of carcinogenicity in polycyclic aromatic hydrocarbons: Use in short-term tests. In: Cooke M, Dennis AJ (eds), Polynuclear aromatic hydrocarbons: Chemistry characterisation and carcinogenesis. Battelle Press, Columbus, OH, pp 839-854.
Shields PG, Sugimura H, Caporasso NE, Petruzzelli SF, Bowman ED, Trump BF, Weston A, Harris CC (1992) Polycyclic aromatic hydrocarbon-DNA adducts and the CYP1A1 restriction fragment length polymorphism. Environ Hlth Perspect 98:191-194.
Shugart L (1985) Quantitating exposure to chemical carcinogens. *In vivo* alkylation of haemoglobin by benzo[*a*]pyrene. Toxicology 34:211-220.
Silverman BD, Lowe JP (1982) Carcinogenicity of methylated hydrocarbons: Effect

of methylation on the calculated diol-epoxide reactivity. Cancer Biochem Biophys 6:89-91.
Silverman BD, Lowe JP (1986) Calculated nitrenium ion stability of carcinogenic/mutagenic aromatic amines. In: Cooke M, Dennis AJ (eds), Polynuclear aromatic hydrocarbon chemistry, characterization and carcinogenesis. Battelle Press, Columbus, OH, pp 855-863.
Silverman BD, Lowe JP (1988) Calculated reactivities of the ultimate carcinogens of polycyclic aromatic hydrocarbons (PAH):A useful tool in predicting structure-activity relationships. In: Yang SR, Silverman BD (eds), Polycyclic aromatic hydrocarbon carcinogenesis: Structure-activity relationships, vol 2. CRC Press, Boca Raton, FL, pp 89-109.
Sims P, Grover PL (1974) Epoxides in polycyclic aromatic hydrocarbon metabolism and carcinogenesis. Adv Cancer Res 20:165-274.
Sims P, Grover PL (1981) Involvement of dihydrodiols and diol-epoxides in the metabolic activation of polycyclic hydrocarbons other than benzo[*a*]pyrene. In: Gelboin HV, Ts'O POP (eds), Polycyclic hydrocarbons and cancer, vol 3. Academic Press, New York, pp 117-181.
Skipper PC, Green LC, Bryant MS, Tannenbaum SR (1984) Monitoring exposure to 4-aminobiphenyl via blood protein adducts. Berlin A, Draper M, Hemminki K, Vainio H (eds) Monitoring human exposure to carcinogenic and mutagenic agents. IARC Scientific Publ 59:143-150.
Smith IA, Berger GD, Seybold PG, Serve MP (1978) Relationships between carcinogenicity and theoretical reactivity indices in polycyclic aromatic hydrocarbons. Cancer Res 38:2968-2977.
Springer DL, Mahlum DD, Westerberg RB, Hopkins KL, Frazier ME, Later DW, Weimer WC (1986) Carcinogenesis, metabolism and DNA binding studies of complex organic mixtures. In: Cooke M, Dennis AJ (eds), Polynuclear aromatic hydrocarbons, chemistry, characterization and carcinogenesis. Battelle Press, Columbus, OH, pp 881-891.
Stewart BW (1992) Role of DNA repair in carcinogenesis. In: Vainio H, Magel PN, McGregor DB, McMichael AJ (eds), Mechanisms of carcinogenesis in risk identification. Internat Agency for Res on Cancer, Lyon, France, pp 307-320.
Stillwell WG, Bryant MS, Wishnok JS (1987) GC/MS analyses of biologically important aromatic amines. Application to human dosimetry. Biomed Environ Mass Spectrom 14:221-223.
Stowers SJ, Anderson MW (1985) Formation and persistence of benzo[*a*]pyrene metabolite-DNA adducts. Environ Hlth Perspect 62:31-39.
Stowers SJ, Maronpot RR, Reynolds SH, Anderson MW (1987) The role of carcinogens in chemical carcinogenesis. Environ Hlth Perspect 75:81-86.
Sugimura T, Terada M, Yokota J, Hirohashi S, Wakabayashi K (1992) Multiple genetic alterations in human carcinogenesis. Environ Hlth Perspect 98:5-12.
Swenberg JA, Richardson FC, Boucheron JA, Dryoff MC (1985) Relationship between DNA adduct formation and carcinogenesis. Environ Hlth Perspect 62:177-183.
Szentpaly LV (1984) Carcinogenesis by polycyclic aromatic hydrocarbons: A multilinear regression on new type P.M.O. indices. J Am Chem Soc 106:6021-6028.
Taningher M, Saccomanno A, Santi L, Grilli S, Parodi S (1990) Quantitative pre-

dictability of carcinogenicity of the covalent binding index of chemicals to DNA: Comparison of the *in vivo* and *in vitro* assays. Environ Hlth Perspect 84:183–192.

Takematsu M, Nagamine Y, Farber E (1983) Redifferentiation as a basis for remodeling of carcinogen-induced hepatocyte nodules to normal appearing liver. Cancer Res 43:4049–4058.

Thomas H, Schladt L, Doehmer J, Knehr M, Oesch F (1990) Rat and human liver cytosolic epoxide hydrolases: Evidence for multiple forms at the level of protein and mRNA. Environ Hlth Perspect 88:49–56.

Tierney B, Benson A, Garner RC (1986) Immunoaffinity chromatography of carcinogen DNA adducts with polyclonal antibodies directed against benzo[a]pyrene diol epoxide-DNA. J Natl Cancer Inst 77:261–267.

Tornqvist M, Kautiainen A (1993) Adducted proteins for identification of endogenous electrophiles. Environ Hlth Perspect 99:39–44.

Vaca CE, Lofgren M, Hemminki K (1992) Some quantitative considerations about DNA adduct enrichment procedures for ^{32}P-postlabelling. Carcinogenesis 13:2463–2466.

Vahakangas K, Yrjanheikki E (1990) Synchronous fluorescence spectrophotometry of benzo[a]pyrene diol epoxide-DNA adducts in workers exposed to polycyclic aromatic hydrocarbons. In: Vainio H, Sorsa M, McMichael AJ (eds), Complex mixtures and cancer risk. IARC scientific publ no 104, Internat Agency for Res on Cancer, Lyon, France, pp 199–204.

Varanasi U, Reichert WL, Stein JE (1989) ^{32}P-postlabelling analysis of DNA adducts in liver of wild english sole (*Parophys vectulus*) and water flounder (*Pseudopleuronectes americanus*). Cancer Res 49:1171–1177.

Walker CL (1989) Oncogenes and cancer suppression genes: Components in a multistage model of carcinogenesis. CIIT Activities 9:1–5.

Wang R, O'Laughlin JW (1992) Determination of DNA-benzo[a]pyrene adducts by high-performance liquid chromatography with laser-induced fluorescence detection. Environ Sci Technol 26:2294–2297.

Weston A, Manchester DK, Poirier MC, Choi JS, Trivers GE, Mann DL, Harris CC (1989a) Derivative fluorescence spectral analysis of polycyclic aromatic hydrocarbon-DNA adducts in human placenta. Chem Res Toxicol 2:104–108.

Weston A, Rowe ML, Manchester DK, Farrer PB, Mann DL, Harris CC (1989b) Fluorescence and mass spectral evidence for the formation of benzo[a]pyrene *anti*-diol-epoxide-DNA and -haemoglobin adducts in humans. Carcinogenesis 10:251–257.

Weston A, Bowman ED (1991) Fluorescence detection of benzo[a]pyrene-DNA adducts in human lung. Carcinogenesis 12:1445–1449.

Weyand EH, Bryla P, Wu Y, He ZM, La Voie EJ (1993) Detection of the major DNA adducts of benzo(*j*)fluoranthene in the mouse skin: Nonclassical dihydrodiol epoxides. Chem Res Toxicol 6:117–124.

Whong WZ, Stewart JD, Org T (1992) Comparison of DNA adduct detection between two enhancement methods of the ^{32}P-postlabelling assay in rat lung cells. Mutat Res 283:1–6.

Wild D (1990) A novel pathway to the ultimate mutagens of aromatic amino and nitro compounds. Environ Hlth Perspect 88:27–31.

Wilson BW, Later DW, Haugen DA (1988) Chemistry of coal conversion materials related to toxicology and process. In: Gray RH, Drucker H, Massey JJ (eds), Toxicology of coal conversion processing. Wiley, New York, pp 371-383.

Wislocki PG, Lu AYH (1988) Carcinogenicity and mutagenicity of proximate and ultimate carcinogens of polycyclic aromatic hydrocarbons. In: Yang SK, Silverman BD (eds), Polycyclic aromatic hydrocarbon carcinogenesis: Structure-activity relationships, vol 1. CRC Press, Boca Raton, FL, pp 1-30.

Wogan GN, Gorelick NJ (1985) Chemical and biological dosimetry of exposure to genotoxic chemicals. Environ Hlth Perspect 62:5-18.

Wolff RK, Bond JA, Sun JD, Henderson RF, Harkema JR, Griffith WC, Mauderly JL, McClellan RO (1989) Effects of adsorption of benzy(a)pyrene onto carbon black particles on levels of DNA adducts in lungs of rats exposed by inhalation. Toxicol Appl Pharmacol 97:289-299.

Wornat MJ, Braun AG, Hawiger A, Longwell JP, Sarofim AF (1990) The relationship between mutagenicity and chemical composition of polycyclic aromatic compounds from coal pyrolysis. Environ Hlth Perspect 84:193-201.

Yang SK, Chou MW, Fu PP (1980) Metabolic and structural requirements for the carcinogenic potencies of unsubstituted and methyl-substituted polycyclic aromatic hydrocarbons. In: Pullman B, Ts'O POP, Gelboin H (eds), Carcinogenesis: Fundamental mechanisms and environmental effects. D. Reidel Publishing, Dordrecht, The Netherlands, pp 143-154.

Yang SK (1988) Metabolism and activation of benz[a]anthracene and methylbenz[a]anthracenes. In: Yang SK, Silverman BD (eds), Polycyclic aromatic hydrocarbon carcinogenesis: Structure-activity relationships, vol 1. CRC Press, Boca Raton, FL, pp 129-149.

Ying TS, Enomoto K, Sarma DSR, Fowler E (1982) Effects of delays in the cell cycle on the induction of preneoplastic and neoplastic lesions in rat liver by 1,2 dimethylhydrazine. Cancer Res 42:876-880.

Zacharias DE, Kashino S, Glusker JP, Harvey RG, Amin S, Hecht SS (1984) Bay-region geometry of sore 5-methylchrysenes: Steric effects in 5,6- and 5,12-dimethylchrysenes. Carcinogenesis 5:1421-1424.

Manuscript received August 25, 1993; accepted September 1, 1993.

Interactions of Pesticides and Metal Ions with Soils: Unifying Concepts

Donald S. Gamble,* Cooper H. Langford,† and
G.R. Barrie Webster‡

Contents

I. Introduction	63
II. Theory and Strategic Concepts	65
A. Research Strategy	65
B. Categories of Variables	67
C. Mathematical Descriptions of Equilibria	68
D. Mathematical Descriptions of Kinetics	70
III. Analytical Chemistry	72
A. Stoichiometry in Mixed Geochemical Systems	72
B. Chemical Speciation Methods	77
IV. Experimental Results	79
V. Success of Research Strategy: Experimental Tests	82
A. Weighted-Average Equilibrium Functions	82
B. Chemical Equilibria in Mixed Geochemical Systems	83
C. Acid Catalysis of Hydrolysis in Fulvic Acid	83
D. Evidence for Generality in Mixed Geochemical Systems	84
E. Intraparticle Diffusion	85
F. Confirming Laboratory Results in the Field	87
Summary	89
References	89

I. Introduction

Thirty years ago, it was universally assumed that it would never be possible to do rigorously quantitative chemistry with such complicated natural mixtures as agricultural soils, aquatic sediments, natural waters, or even their components. One of the practical implications of that assumption was that the physical chemistry interactions of metal ions and organic chemicals with the natural geochemical mixtures could never be understood or described quantitatively in terms of molecular-level mechanisms. Practical chemical problems arising from the use of fertilizers and pesticides were consequently

---

*Centre for Land and Biological Resource Research, Agriculture Canada, Research Branch, Ottawa, Ontario, Canada K1A 0C6.

†Department of Chemistry, University of Calgary, 2500 University Drive, Calgary, Alberta, Canada T2N 1N4.

‡Department of Soil Science, University of Manitoba, Winnipeg, Manitoba, Canada R3T 2N2.

© 1994 by Springer-Verlag New York, Inc.
Reviews of Environmental Contamination and Toxicology, Vol. 135.

formulated in empirical or semiempirical forms. This approach led to an important practical limitation in soil chemistry and environmental chemistry.

Empirical descriptions represent only the particular experiments out of which they have come and do not support reliable predictions. The types of extrapolations and generalizations that make it possible to compare different samples or experiments, and to do predictive calculations, for instance, for practical hydrology engineering, require exact mathematical descriptions of real processes or mechanisms. For such physical chemistry interactions as labile sorption and desorption, retarded intraparticle diffusion, and catalyzed chemical reactions, the mathematical descriptions must be based on chemical stoichiometry.

There is now a general consensus that research on geochemical systems and environmental problems must become much less empirical (Buffle 1988; Buffle and Altmann 1987; Langford and Gutzman 1992; Shuman et al. 1983). Classical physical and analytical chemistry can now be done in this type of research, and the research strategy that has evolved permits the use of chemical stoichiometry. The mechanism by which the strategy has evolved has consisted of a sequence of investigations that have followed a logical progression through a set of geochemical systems of increasing complexity.

This increasing complexity continues to move toward practical field conditions. In the first stage, many authors published physical and chemical properties of clays (Grim 1953; Newman 1987), humic materials (Aiken et al. 1985), and hydrous metal oxides (Taylor 1987). The second stage consisted of investigations into the solution phase reactions of fulvic acid with cations (Gamble 1972, Burch et al. 1978; Gamble et al. 1980; Gamble and Langford 1988; Gamble et al. 1985) and organic pesticides (Gamble et al. 1986; Wang et al. 1991). This led automatically to the work on cation and pesticide reactions with undissolved humic acid in two physical phases (Gamble and Khan 1988; Sojo 1992; Gamble and Khan 1990). In the fourth stage of the progression, the work with humic acid in two physical phases is providing both theoretical and experimental support for work with whole mineral soils (Gamble and Khan 1992; Gamble and Khan 1985; Gamble et al. 1983; Bowman 1990). Research with aquatic sediments also belongs to this fourth stage of complexity. At this level, the laboratory investigations can begin to aspire to the realism of field conditions. Progress can be determined by entering laboratory studies of mechanisms into hydrology computer models and then testing the computer predictions against field experiments.

Since the research strategy is not yet very widespread and its effects are only now emerging, its nature and validity warrant a careful examination. Other authors have published excellent reviews of the mathematical methods for interpreting equilibria and kinetics in mixed geochemical systems (Buffle 1988; Buffle and Altmann 1987; Langford and Gutzman 1992;

Shuman et al. 1983). The next step is to show that the whole research strategy is now yielding some successes by referring to published and current research.

II. Theory and Strategic Concepts

The creation and application of the concepts, mathematical descriptions, and to some extent the experimental methods have been guided largely by the research strategy. They are therefore best seen within its context. Although most of its components are separately familiar and obvious, their assembly into a coherently integrated research strategy is an achievement that is making an important impact. Stoichiometrically exact chemistry can now be done with complex geochemical mixtures such as soils, sediments, and natural waters in the same sense that exact chemistry is done with monomeric pure reagents. It is possible to produce mathematical descriptions of physical chemical mechanisms that can be used for predictive calculations relating to real-world conditions. Several components of the research strategy can be itemized.

A. Research Strategy

1. Mathematical Descriptions: Three Types of Information

a. Constants and variables that can be theoretically deduced. Examples might include the number of bidentate Cu(II) chelation sites of a humic material, or the pesticide sorption capacity of a mineral soil.
b. Constants or variables that provide the required useful answers. These include the equilibrium constants and kinetic rate constants that characterize a mechanism, mole fractions of occupied binding sites, and predicted solution concentrations.
c. The mathematical equations that state the behavior and constraints of the system under investigation. They will typically provide an integrated equilibrium-kinetics description of such mechanisms as the multiple cation exchange of a humic acid or acid-catalyzed hydrolysis of a pesticide in a mineral soil.

The reason why a mechanism should be formulated in this way is that the calculation process is simply the means by which it is possible to go from the measurements that it is possible to make to the answers that it is necessary or important to use.

The design and construction of a bridge provide an analogy. In this case, the three types of information include:

a. The properties of the construction materials and load-bearing properties of the building site
b. The performance requirements that the completed bridge must meet

c. The mathematical equations that describe the constraints imposed on the design by the properties of the materials, load-bearing capacity of the bridge site, and performance requirements

2. Mixtures Treated as Chemical Systems. Mixtures of nonidentical reactive sites or binding sites are treated as chemical systems, rather than as large numbers of separate components.

3. Weighted Averages, the Law of Mass Action, and Chemical Kinetics. The concept of weighted averages is applied to the law of mass action, for equilibria in the natural mixtures of geochemical systems. A different form of the concept has been applied to chemical kinetics in the mixed systems.

4. Using Categories of Variables in Experimental Design and Interpretation. The characteristics of different categories of variables are used for the design of experiments, interpretive and predictive calculations, and reporting of research results.

5. Categories of Reactive Sites. Categories of reactive or binding sites are identified when possible.

6. Units Prerequisite for Chemical Stoichiometry. Chemical units such as moles or (mol/L) are used for mass and concentration instead of mg or ppm, in the quantitative descriptions of material balances, equilibria, and chemical kinetics. This is a prerequisite to the application of chemical stoichiometry.

7. Sorption Capacities and Stoichiometry. Total numbers of reactive or binding sites are measured as capacities or saturation limits. The resulting numerical values are used in defining the stoichiometries of equilibrium and kinetics calculations.

8. Characterization of Mixed Geochemical Systems by Titration. Natural mixed systems are chemically characterized in terms of particular chemical reactions or binding processes, by means of experimental scans across the components of the mixtures, for example, by some kind of titration.

9. Concise Experimental Methods for Quantitative Descriptions of Equilibria and Kinetics. Experimental methods have been developed with which free, labile-bound, and relatively less labile-bound chemical species may be identified and tracked kinetically. This permits mechanisms to be quantitatively described.

B. Categories of Variables

Three categories of variables should be clearly distinguished from each other, for the conduct, calculation, and molecular-level interpretation of experiments with these environmental chemistry systems (Gamble and Khan 1992). They are described below.

1. Inner Variables. Inner variables are those identified by theory as being required for totally specifying the state of a chemical system. They should be used for correlations, molecular-level interpretative calculations, and predictive calculations. Typical examples include the numbers or concentrations of protonated carboxyl groups and the mole fraction of sorption sites occupied by a sorbed pesticide.

2. Outer Variables. Outer variables are the type that are manipulated operationally for the conduct of an experiment. Typical examples include pH, total moles of pesticide or metal ion added to a sample, and the ratio of weight of solid sample to volume of experimental solution. They are more practical than inner variables for setting up and running many types of experiments, and frequently inner variables cannot be used for that purpose.

3. Background Descriptive Variables. These variables are useful as general background information for developing qualitative insight or making practical decisions. They are not used in chemical calculations and do not appear in the mathematical descriptions of the system. Typical examples include elemental analysis and percent organic matter in a mineral soil.

Incorrect use of the three types of variables can cause unnecessary data scatter and some confusion. For example, hydrogen bond structure formation (Wang et al. 1991; Gamble and Khan 1985) may cause a sorption rate constant to be a function of the mole fraction of the sorption sites covered by pesticide molecules. A plot of the rate constants for several soils against pH would likely show excessive data scatter. The reason is that, at one and the same pH, different soils would have different numbers of protonated carboxyl groups and therefore different mole fractions of their sites covered. The outer variable pH would be no more than a pale, indirect reflection of the correct inner variable, the mole fraction of the sorption sites covered. If in this case pH were used for empirical black-box correlations, the resulting fitted curve would represent only the particular experiment from which it had come. Because it would not represent the actual molecular-level phenomena, it would lack the generality necessary for good predictive calculations. For the above reasons, it is important to compare experimentally measured dependent variables to the proper independent variable or at least be aware of the danger of using a variable from the wrong category.

C. Mathematical Descriptions of Equilibria

It has been known for more than twenty years that equilibria in mixed geochemical systems can be described by weighted-average functions (Buffle 1988; Gamble and Khan 1992). Equation (1) gives the general form in which $Y(\chi)$ is averaged, with $(\Delta\chi)_i$ being the weighting factors

$$\overline{Y}(\chi) = \left(\sum_{i=1}^{n} (\Delta\chi)_i Y_i\right) \Big/ \left(\sum_{i=n}^{n} (\Delta\chi)_i\right) \tag{1}$$

Two formulations of the problem have been reported in the literature and extensively reviewed (Buffle 1988; Shuman et al. 1983). The earliest one was based on the macroscopically measured equilibrium function K expressed as a function of the extent of reaction in Eq. (2), in which χ is the mole fraction of unconverted reactant

$$\overline{K} = f_k(\chi_1) \tag{2}$$

With $\overline{Y} = \overline{K}$, the application of Eq. (1) to this gave Eq. (3), in which the equilibrium constants for components of the mixture were averaged, and the weighting factors were mole fractions of unreacted components in the mixture

$$\overline{K} = \left(\sum_{i=1}^{n} K_i(\Delta\chi)_i\right) \Big/ \left(\sum_{i=1}^{n} (\Delta\chi)_i\right) \tag{3}$$

In its most advanced form, this was extended by the introduction of a material balance relationship to give a theoretical description of a multiple metal ion-exchange system (Gamble et al. 1983). The resulting Eq. (4) is a general working equation that relates the weighted-average equilibrium function for a given metal ion "M" to the fraction of sites covered by it, χ_{c1}, and to the effects of all of the other metal ions

$$\overline{K}_M = -\frac{1}{\chi_{SH2}} \int_0^{\chi_{c1}} K_M \, d\chi_{c1} - \sum_{J=2}^{k} \frac{1}{\chi_{SH2}} \int_0^{\chi_{cj}} K_j d\chi_{cj} \tag{4}$$

The terms under the summation account for the other metal ions. χ_{SH2} is the mole fraction of exchange sites occupied by protons. One such equation may be written for each of the metal ions in the system. An important point is that even for 11 metal ions competing for a mixture of nonidentical ion-exchange sites, all of the chemical composition variables were accounted for. They are, in fact, the inner variables which totally specify the chemical state of the system. No other formulation of the mixed equilibria problem has yet achieved this. It should be noted that pH does not appear in the general working equation in spite of the fact that carboxylate anions contribute ion-exchange sites. The reason is that χ_{SH2} is the relevant inner variable accounting for proton loading of the carboxyls. pH is consequently an outer variable which should not be used for correlations or predictive calculations. Subject to some limitations, the averaging inherent in the $\overline{K}$ functions can be undone by simple first derivatives or partial derivatives to give differential equilibrium functions (Buffle 1988; Gamble et al. 1980; Gamble and Langford 1988; Gamble et al. 1983). The purpose of this is to permit molecular-level effects to be observed free from the distortions of the averaging process.

After Eq. (3) was introduced, Klotz and Hunston (1984) applied the weighted-average concept of Eq. (1) to the inverse function in Eq. (5) (Gamble and Langford 1988)

$$\overline{\chi}_1 = f_\chi(K) \tag{5}$$

The mole fraction of reacted sites is averaged in this formulation of the problem, with component equilibrium constants becoming the weighting factors, as in Eq. (6)

$$\overline{\chi}_1 = \left(\sum_{i=1}^{n} \chi_i K_i\right) \bigg/ \left(\sum_{i=1}^{n} K_i\right) \tag{6}$$

Assuming a continuous distribution $\chi_1 = \chi_1(K)$, the summations can be replaced by integrations. Mathematical transformations put the weighting factor into the form $f(\ln K)\, d\ln K$.

With the symbols used by Nederlof, van Reimsdyk, and Koopal (Nederlof et al. 1990; Nederlof 1992; de Wit 1992), this is usually in the form of Eq. (7).

$$\Theta = \int_0^\infty \theta f(\ln K)\, d\ln K \tag{7}$$

Θ is the mole fraction of the total binding sites that are occupied. Θ is the continuous function corresponding to Θ_{si}, the mole fraction of the ith component that is occupied. In the same way, the distribution function $f(\ln K)$ corresponds to (C_{Ti}/C_T), the mole fraction of the total ith component in the whole mixture of binding sites. Also, $\mathcal{K} = KM_A^0$, in which K is the continuous function related to the equilibrium constant K_i for the ith component. M_A^0 is the mol/L of total (free + bound) reactant in the system. Θ is given by the Langmuir isotherm, Eq. (8)

$$\Theta = (\mathcal{K}R)/(1 + R) \tag{8}$$

$R = (M_A/M_A^0)$ is the mole fraction of the total reactant remaining free in solution. The origins of this description can be traced back through more than forty years of surface chemistry research (Nederlof et al. 1990; Nederlof 1992; de Wit 1992; Harris 1968). The problem addressed by numerous authors is how to extract the distribution function $f(\ln \mathcal{K})$ from the integral in Eq. (7)

The LOGA method presented by Nederlof et al. (1990) for the extraction of $f(\ln \mathcal{K})$ carries out the calculation in five steps. In the first step, Eq. (8) is replaced by the approximations in Eqs. (9)

$$Z_1 = \alpha(\mathcal{K}\chi)^\beta \qquad (9a)$$

$$Z_2 = 1 - \alpha(\mathcal{K}\chi)^{-\beta} \qquad (9b)$$

The second step adapts the Scatchard plot concept to the mixture of components. For the ith component, this gives Eq. (10), with $\Theta_i = 0.5$

$$\mathcal{K}_i\chi_i = \frac{\Theta_i}{1 - \Theta_i} = 1 \qquad (10)$$

The subscripts are dropped when summation is replaced by integration. In the third step, the integral in Eq. (7) is split at the limit $\ln \mathcal{K}^*$, which is defined by $\mathcal{K}\chi = \mathcal{K}^*\chi^*$ at $\Theta^* = 0.5$. These changes give Eq. (11)

$$\Theta(\chi^*) = \frac{1}{2\mathcal{K}^*} \int_0^{\ln \mathcal{K}^*} \mathcal{K} f(\ln \mathcal{K}) \, d\ln \mathcal{K} + \int_{\ln \mathcal{K}^*}^\infty f(\ln \mathcal{K}) \, d\ln \mathcal{K} \qquad (11)$$

$$- \frac{\mathcal{K}^*}{2} \int_{\ln \mathcal{K}^*}^\infty \frac{1}{\mathcal{K}} f(\ln \mathcal{K}) \, d\ln \mathcal{K}$$

Step four uses a combination of partial differentiation and subtraction of terms with which to recover the distribution function from the integrals. Practical calculations then become possible with the resulting Eq. (12)

$$f_{LOGA}(\ln \mathcal{K}) = \frac{\partial \Theta(\chi)}{\partial \ln \chi} - \frac{0.189}{\beta^2} \frac{\partial^3 \Theta(\chi)}{\partial (\ln \chi)^3} \qquad (12)$$

In the final step, experimental data are fitted by the method of least squares to curves of Θ versus $\ln \chi$. Eq. (12) then uses the partial derivatives with $\beta = 0.07$. It should be noted that the mole fraction $f(\ln K) = (c_T/C_T)$ used here is not the same as the mole fraction $\Theta_0 = (C_0/C_T)$, with which the differential equilibrium function calculations are carried out.

D. Mathematical Descriptions of Kinetics

The same problem of mixtures of reactive sites is encountered in the reaction kinetics of mixed geochemical systems. The Shuman kinetics (Shuman et al. 1983) and Langford kinetics (Langford and Gutzman 1992) have both used inverse Laplace transform calculations for extracting rate-constant distributions from macroscopic kinetics measurements on the whole mixtures. The ith component of the whole mixture has an initial concentration of C_{0i}, and $C(t)$ is the total concentration of all of the unreacted components at a subsequent time t. A common reagent, which could be a small cation or organic chemical molecule, reacts with each of the components to give a set of similar nonidentical reaction rate laws. The method is restricted to systems having first-order or pseudo-first-order kinetics behavior. The macroscopically measured total kinetics effect is then described by the sum of all of the components terms in Eq. (13)

$$C(t) = \sum_{i=1}^{n} C_{0i} e^{-k_i t} \tag{13}$$

The k_i terms are component rate constants. The reverse Laplace transform calculation is based on Eq. (14), in which the use of the symbol $C(k, t)$ recognizes the fact that the concentration of unreacted mixture depends both on time and the numerical values of component rate constants

$$C(k, t) = S \int_{-\infty}^{\infty} H(k, t) e^{-kt} d(\ln k) \tag{14}$$

The factor S is an experimentally determined function of time, and the function $H(k, t)$ representing the frequencies of component $\ln k$ values is explained by Eq. (15)

$$H(k, t) = \frac{\partial^2 C(k, t_x)}{\partial (\ln t)^2} - \frac{\partial C(k, t)}{\partial (\ln t)} \tag{15}$$

Practical calculations use Eq. (15) with numerical methods for the evaluation of $H(k, t)$. Eq. (15) is, in fact, only an approximation that has been obtained by truncating a series of increasingly higher partial derivatives. It is used as a practical compromise.

Rate-constant distribution functions are obtained from graphical plots of $H(k, t)$ versus $\ln (2/t)$. The argument $(2/t)$ has the numerical value of the rate constant k. If a finite number of discrete frequency peaks can be found in this way, then this implies a finite number of component rate constants. This, in turn, implies a finite number of generic types of chemical components in the reacting mixture. The frequency-function peaks generally have finite widths. There are at least two possible reasons. The most fundamental reason is that a complicated natural mixture might have generic types of components, each of which is itself a mixture. The other is that the practical calculations themselves have a broadening effect.

The Langford kinetics analysis has introduced an important strategy. They used the inverse Laplace transform calculations to determine the number of component rate constants and then used that number of parameters for least-squares fits of the experimental data. The least-squares fit procedure can give better numerical values for the rate constants if it is known how many of them there are. Unlike the inverse Laplace transform calculation, a least-squares fit cannot determine objectively how many of them exist.

Since mathematical artifacts remain a danger in all numerical differentiation procedures and a finite number of components may generate only $f(k)$ parameters, Langford's group has also insisted on chemical consistency tests.

All of these theoretical analyses for both equilibria and kinetics have several features in common. They all begin by recognizing that macroscopic measurements of equilibrium or kinetic rate constants made on mixtures will produce average values. They must generally be some sort of weighted averages of the values for components. These macroscopic values will probably be the most useful for modeling and predicting field processes. At the same time, the elucidation of molecular-level mechanisms will require some kind of information related to the numerical values of component constants. The practical importance of molecular-level mechanisms is that they make predictive calculations possible. This, in turn, opens up new opportunities for the proper management of field operations. All of the theoretical schemes consequently use some sort of differentiation that attempts to "undo the averaging" that happens when macroscopic measurements are made on whole mixtures. This includes, for example, the differential equilibrium function method and LOGA method (Nederlof et al. 1990). All of them require that care be taken to prevent mathematical artifacts in the practical calculations.

III. Analytical Chemistry
A. Stoichiometry in Mixed Geochemical Systems

In the reaction of a monomeric reagent with a mixed geochemical system, the stoichiometry is governed by the numbers of molecules, ions, and binding or reactive sites, and the usual rules of chemical valence. This differs in two ways from the reaction between two pure monomeric reagents. The first is that the reactants include binding or reactive sites that do not have an independent existence. Instead, they exist on organic polymers, inorganic particles of colloidal dimensions or larger, or some combination of them. The second is that generally the binding or reactive sites are mixtures, whose components might be chemically similar but not identical. It is important to note that this does not prevent the measurement of their total numbers. In some cases, especially with protons or cations, the binding or reactive sites can be identified. Carboxylate anions, bidentate chelation sites, and the cation-exchange sites of clays are common examples. There are other cases, however, that are more subtle, more complicated in their behavior and less readily identified. Such cases include the solution phase complexing of organic pesticides by hydrogen bonding to fulvic acid and the sorption of organic pesticides at the water–solid interface of immersed soil particles. In spite of these apparent problems, analytical chemical methods have been developed for measuring the total numbers of such binding and reactive sites. This is the key to the stoichiometry in the reaction of a small ion or molecule with a mixed geochemical system.

1. Dissolved Fulvic Acid: A Weak Acid Polyelectrolyte. Most of the analytical chemical methods produce either titration endpoints, or the plateaus

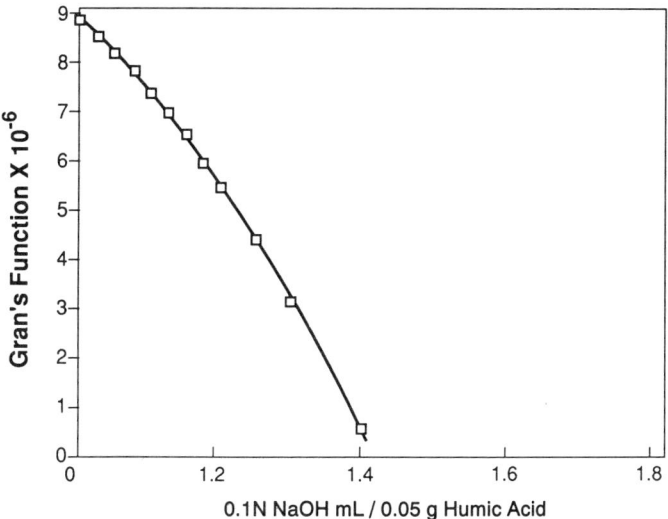

Fig. 1. Equivalence point calculation for humic acid titration with standard NaOH.

of site saturation curves. The simplest physical system proved to be one of the more difficult analytical chemistry problems to solve (Gamble 1972; Burch et al. 1978). Dissolved fulvic acid is a weak acid polyelectrolyte. The spectrum of weak acid constants and the electrostatic work term both contribute to the strong buffering. This interferes with conventional titration endpoint determination. The problem was solved by correcting the Gran's function for the buffering effect, as indicated in Eqs. (16) and (17)

$$Y_1 = (Vm_H - Q_1) = \overline{K}_A(V_{el} - V) \tag{16}$$

$$Q_1 = \frac{(W_0 + R_w R_v V)}{N_B}\left(m_B - m_H + \frac{K_w}{m_H}\right)(\overline{K}_A + m_H) \tag{17}$$

Y_1 is the Gran's function, V is the volume of standard base, V_{el} is the equivalence point value of V, $\overline{K}_A$ is the weighted-average equilibrium function for acid dissociation, Q_1 is the correction term, W_0 is the initial sample weight, N is the normality of the standard base, and m_B and m_H are the molalities of total base added and protons left. R_w and R_v convert weights to volumes. The equations require iterative calculations, but have produced titration equivalence points that were verified by compleximetric titrations with metal ions (Gamble 1972; Gamble et al. 1980; Gamble et al. 1985).

2. Solid-Phase Humic Acid and Cu^{2+} in Aqueous Slurry. Figure 1 represents the next step up in complexity of systems, with two physical phases (Gamble 1989). An aqueous slurry of undissolved humic acid was titrated with standard base. Although the system is physically more complicated,

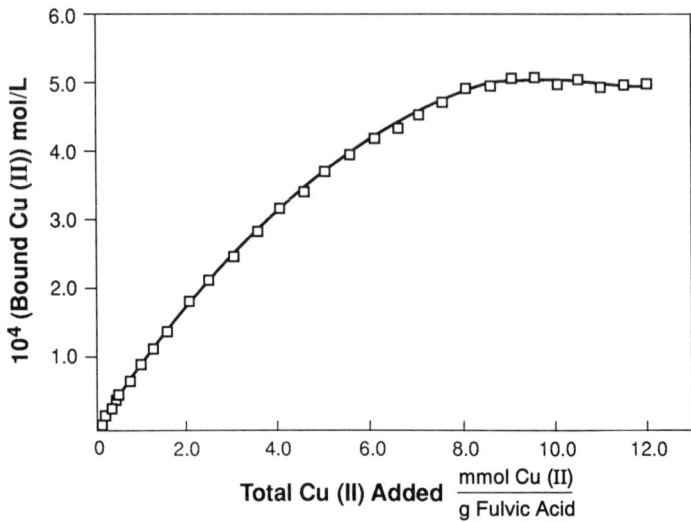

Fig. 2. Complexing capacity of fulvic acid for Cu^{++}. Titration curve plateau gives complexing capacity as a saturation limit.

the titration gave a much simpler equivalence determination than a fulvic acid solution does. The dissolution of anionic polymers produced a classic Gran's plot so that a sharply determined equivalence point was found. In other words, the number of carboxyl groups was quite accurately measured. The compleximetric titration of a fulvic acid solution with Cu^{++} was monitored with a Cu^{++} specific ion electrode, appearing in Fig. 2 (Gamble et al. 1980). The available complexing sites were saturated with Cu^{++}, so that their number was accurately measured.

Ultrafiltration can be used to measure directly the free Cu^{++} in a compleximetric titration of humic material (Wang et al. 1991). The resulting plot in Fig. 3 gives a direct determination of the number of complexing sites.

The Cu^{++} complexing capacities of humic materials measured by these methods should decrease with increasing protonation of the carboxyl groups. This expected effect on the stoichiometry is confirmed by the experimental plot in Fig. 4 (Gamble et al. 1980).

3. Dissolved Fulvic Acid and Atrazine. The complexing capacity of dissolved fulvic acid for atrazine can be just as effectively measured using an ultrafiltration technique (Gamble et al. 1986; Wang et al. 1991). The mechanism that governs the size of this capacity is more complicated, however, than it is in the case of metal ions. Figure 5 shows the saturation of hydrogen-bonding sites on protonated carboxyl groups. The linear relationship in Figure 6 reveals the significant fact that only a fraction of a mole

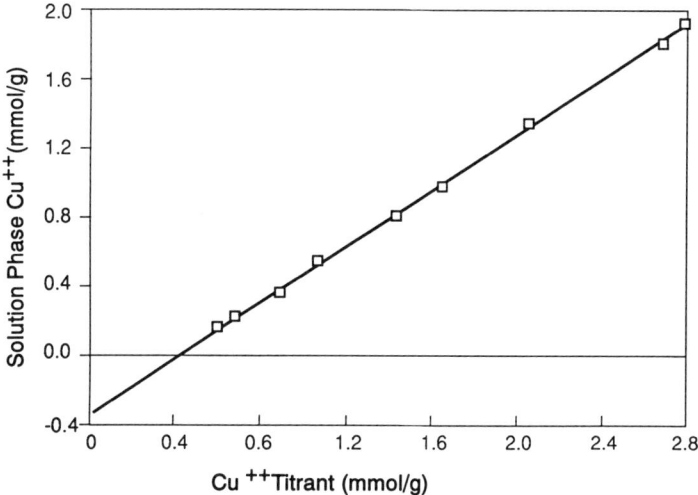

Fig. 3. Cu^{++} complexing capacity of fulvic acid measured by compleximetric titration. Uncomplexed Cu^{++} after site saturation was monitored by ultrafiltration and atomic absorption spectrophotometry.

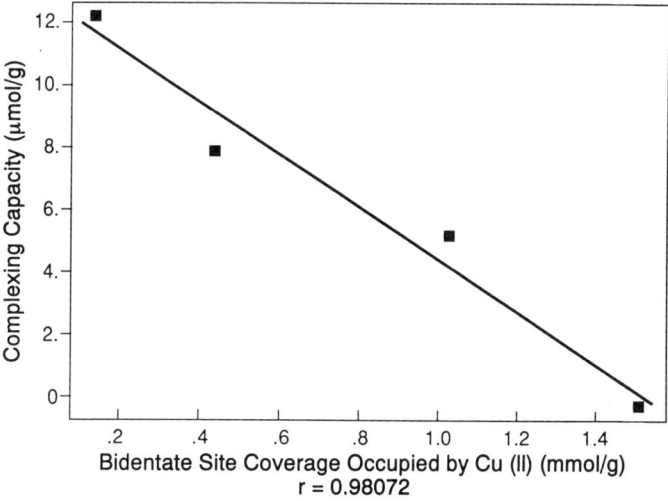

Fig. 4. Interaction of atrazine and Cu^{++} complexing on carboxyl binding sites of a dissolved fulvic acid.

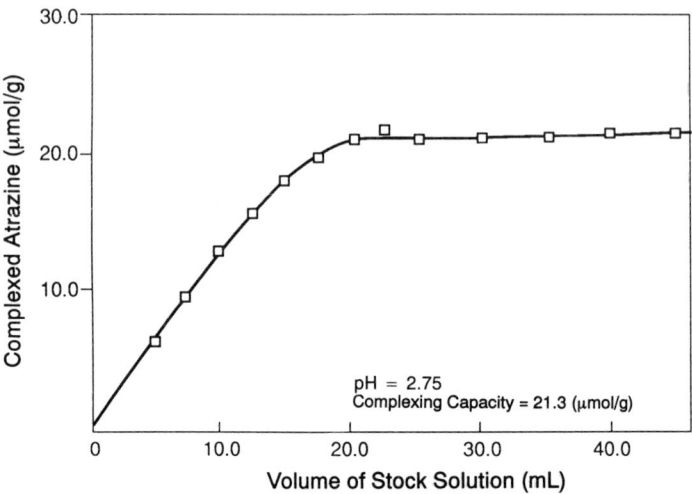

Fig. 5. Atrazine complexing capacity of dissolved fulvic acid. Titration curve plateau gives capacity as a saturation limit. Uncomplexed atrazine was monitored by ultrafiltration and gas chromatography.

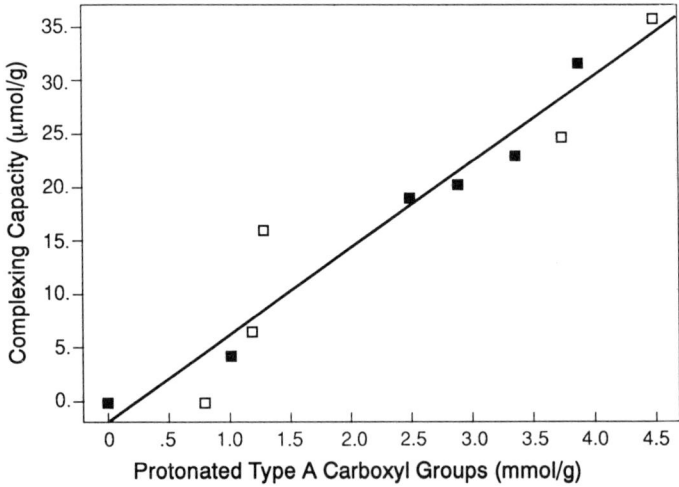

Fig. 6. Relationship of atrazine complexing capacity of dissolved fulvic acid to protonated carboxyl groups concentration. Intercept indicates that complexing vanishes when there are no more carboxyl groups. Slope shows only a fraction of 1% of carboxyl groups participate at any instant in complexing.

percent of those protonated carboxyl groups actually participate in the complexing. Water and the carboxyl groups form an equilibrium collection of hydrogen-bonded structures, and only a small fraction of them can function as binding sites for the relatively hydrophobic atrazine.

4. Extension to Whole Mineral Soil. Whole mineral soils represent the type of system that is not only the most complicated case, but is also the most important for agricultural and environmental problems.

Current research has developed a high-pressure liquid chromatography-microfiltration method that is being used for the measurement of the pesticide sorption capacities of whole mineral soils (Gamble et al. 1986; Gamble and Khan 1990; Gamble and Khan 1992). Defined and measured as site-saturation limits, these capacities define the stoichiometries required for the equilibrium and kinetics calculations in mechanistic studies. A larger database of such capacities will be required before their governing causes can be elucidated.

B. Chemical Speciation Methods

The importance of the chemical speciation of metal ions and organic chemicals in environmental systems has been recognized for decades. In addition, identifying chemical species and determining their behavior are necessary steps for the elucidation of molecular-level mechanisms. Experimental methods are required that will not create laboratory artifacts by distorting the systems that they are supposed to observe or measure. At least three types of experimental methods for avoiding artifacts have been reported.

1. Ultrafiltration. Ultrafiltration can be used for distinguishing between free and complexed small organic molecules and between free and bound metal ions, in aqueous solutions of fulvic acid (Buffle 1988; Gamble et al. 1986; Gamble and Khan 1990). Complexing capacities and equilibrium functions have both been measured in this way.

2. Microfiltration High Pressure Liquid Chromatography. A method that has been introduced more recently combines high-pressure liquid chromatography (HPLC) with microfiltration, for identifying and tracking the kinetics of pesticide species in aqueous slurries of soil or sediment particles (Gamble and Khan 1990; Gamble and Khan 1992). Microfiltrates and whole slurries are alternately injected into the HPLC at measured times. Figure 7 shows a typical experiment, in which three atrazine species were kinetically monitored in a mineral soil slurry. The dissolved atrazine and reversibly sorbed atrazine are in labile equilibrium. The solid curve represents retarded intraparticle diffusion, which is kinetically slow. The chemical-species piechart in Fig. 8 illustrates another example. The material balance loss has been caused by the retarded intraparticle diffusion. These HPLC-

78 D.S. Gamble, C.H. Langford, and G.R.B. Webster

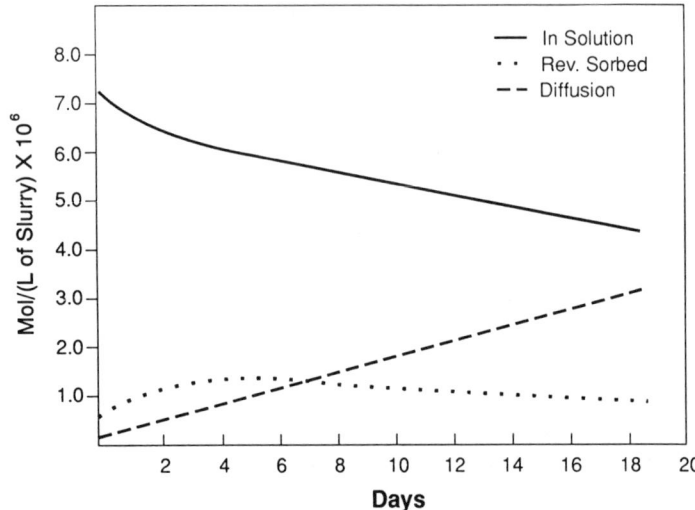

Fig. 7. Behavior of free and bound chemical species of atrazine in aqueous slurry of whole mineral soil at 35.0 °C. Dashed curve shows uptake by intraparticle diffusion.

microfiltration experiments are now being used for the investigation of mechanisms (Gamble et al. 1994). They also have potential applications to toxicology research.

3. Kinetic Rate Constants. The use of kinetic rate constants for characterizing metal ion species in environmental systems has existed for at least

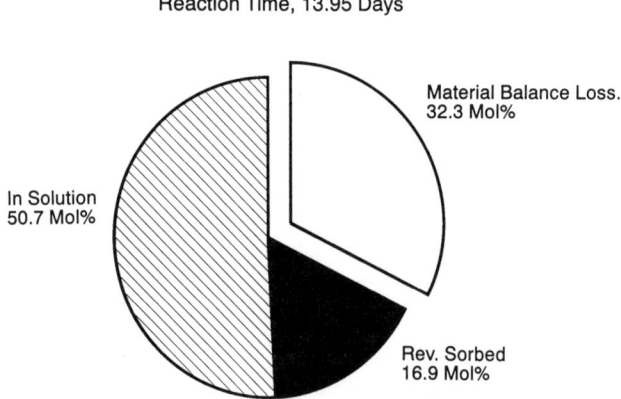

Fig. 8. Typical distribution of atrazine among chemical species in mineral soil slurry. Material balance loss is caused by intraparticle diffusion.

twenty years (Langford and Gutzman 1992). A mixture of complexes of a given metal ion is reacted with a common simple reagent. Langford and various co-workers have refined this strategy by applying the inverse Laplace transform method that was previously mentioned. The component peaks of the rate-constant spectra correspond to the original metal ion species, if it can be confirmed that they display a confirmed pattern of response to chemical variables. The applications of the chemical speciation methods have not yet met their full potential.

IV. Experimental Results

A few experimental results have been selected with which to demonstrate the effectiveness of the research strategy. Buffle (1988) has presented a more extensive review of such results.

One of the earlier examples is shown in Fig. 9. Weak acid spectral curves calculated from titrations by means of the weighted-average equilibrium function theory are given for two fulvic acids (Gamble 1972) that came from two quite different geographical locations.

Although there has been some controversy about the exact chemical meaning of these K_A values, a computer simulation study has established two relevant points (Gamble and Langford 1988). The first is that the K_A values produced by the calculation are more directly related to the molecular-level behavior of the system than is the macroscopically measured weighted-average equilibrium function. They cannot, however, be

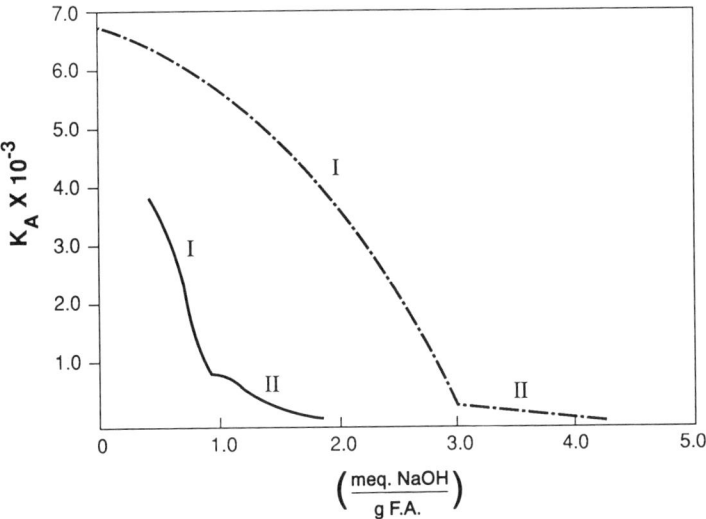

Fig. 9. Differential equilibrium functions for acid dissociation in two fulvic acids. Increasingly weakly acidic carboxyl groups are observed as titration proceeds with standard base.

Table 1. Cu^{++}-Fulvic Acid Chelation Equilibrium at 25.0 °C. Mole Fraction of Fully Protonated Chelating Sites ($\chi_{SH2} \approx 0.37$)

$\overline{K}$ rel. $\sigma = 5\%$	K rel. $\sigma = 20\%$	$-(\Delta G^0 + RT \ln \Gamma)$ $\frac{KJ}{mol} \pm 0.5$	Mol fraction of occupied chelation sites χ_C rel. $\sigma = 2\%$
1.53×10^3	5.9×10^4	27.3	0.0181
1.97×10^2	4.3×10^3	20.7	0.0540
67.5	7.0×10^2	16.2	0.0896

assigned to particular functional groups. The weighted-average equilibrium function can differ from component values by two or more orders of magnitude. The second observation is that some of the component values for the mixture are not recovered by the calculation without some special techniques. An important point about this type of differential equilibrium function is that standard Gibb's energies can reasonably be estimated from them in a way that they cannot from the weighted-average equilibrium functions (Gamble et al. 1980; Gamble and Langford 1988; Gamble et al. 1983). Table 1 illustrates this for Cu^{++} chelation in a fulvic acid solution (Gamble et al. 1980).

Two other details are worth noting. It is seen here that the differential equilibrium function differs from the experimental weighted-average equilibrium function by the usual one to two orders of magnitude. The second point is that the equilibrium functions have been tabulated with the two inner variables that totally specify the chemical state of the system. They are the mole fractions of protonated and occupied binding sites. The customary pH is not an inner variable and has not been listed.

The LOGA calculation method also recovers numerical values for the differential equilibrium function that are more correctly related to molecular-level reactions than the weighted-average equilibrium functions are. It has the additional advantage of minimizing distortions in the relationship of the distribution function $f(\ln K) = (c_T/C_T)$ to the differential K. Two reasons for this are that the calculation relates the differential K to (c_T/C_T) instead of (C_0/C_T), and the Scatchard plot constraint is separately applied to hypothetical components of the whole mixture. A practical limitation is that it has not been adapted to natural systems having several chemical composition variables. The possibility of obtaining exact chemical stoichiometry for the interactions of organic chemical pesticides with humic materials or mineral soils is less obvious than it is for the case of cation reaction with humic materials. The reason for this is that there is no simple relationship between binding or sorption sites and functional groups, as is the case with cations and humic materials. In spite of this, the use of the research

Table 2. Interactions of Atrazine with a Mineral Soil

Surface sorption equilibrium	Intraparticle diffusion kinetics
$\overline{K}_1 = 3.3 \times 10^4$	$k_{D1} = Q\left(\dfrac{D}{a}\right) = 0.131\ \text{d}^{-1}$
$\sigma = 0.9 \times 10^4$	$SE = 0.006\ \text{d}^{-1}$
$\Delta H^0 = 0\ \text{kJ/mol}$	$\Delta E\dagger = 98\ \text{kJ/mol}$
$SE = 10\ \text{kJ/mol}$	$SE = 12\ \text{kJ/mol}$

Gamble and Khan (1990).

strategy has achieved some important successes. An example is represented by Eq. (18), which predicts exactly the half-life of atrazine hydrolysis in an aqueous solution of fulvic acid (Gamble and Khan 1985)

$$t_{\frac{1}{2}} = \frac{8.80 \times 10^{-2}}{(1.78 \times 10^{-2} \times 10^{-\text{pH}})/(2.45 \times 10^{-2} + 10^{-\text{pH}}) + M_{AH}} \text{days} \quad (18)$$

The two variables which together determine the half-life correspond exactly to the two catalysts in the solution that are known to cause the hydrolysis reaction. They are therefore inner variables. The first catalyst is simply the protons free in solution, accounted for by the pH. The other catalyst is the collection of protonated carboxyl groups, for which the total concentration is M_{AH} (mol/L). The concentration of total fulvic acid in (g/L) does not appear in the equation because it is an outer variable. The reason for this is that the bulk fulvic acid polymer is not itself a catalyst. It is instead only a carrier for the two catalysts. This explains why empirical plots of the half-life vs. (g/L) of fulvic acid, with the protons and carboxyl groups being ignored, would be of little use for practical predictions. Depending on what combinations of protons and carboxyl groups were being ignored in such empirical plots, meaningless data scatter could be the result.

The use of labile sorption capacities defined and measured as site saturation limits allows exact chemistry to be done, even with organic pesticides in whole mineral soils. Table 2 lists the resulting information for the labile sorption equilibrium and kinetics of intraparticle diffusion of atrazine in the mineral soil (Gamble et al. 1994). In Eq. (19) for the diffusion rate constant k_{D1}, D is the diffusion coefficient in (m²/s), a the cross-sectional area normal to the direction of diffusion, and Q a conversion factor for time units

$$k_{D1} = Q\left(\frac{D}{a}\right) \quad (19)$$

All of the results in Table 2 are consistent with numerical values reported or found elsewhere (Pignatello 1989). Mineral soil results like those in Table

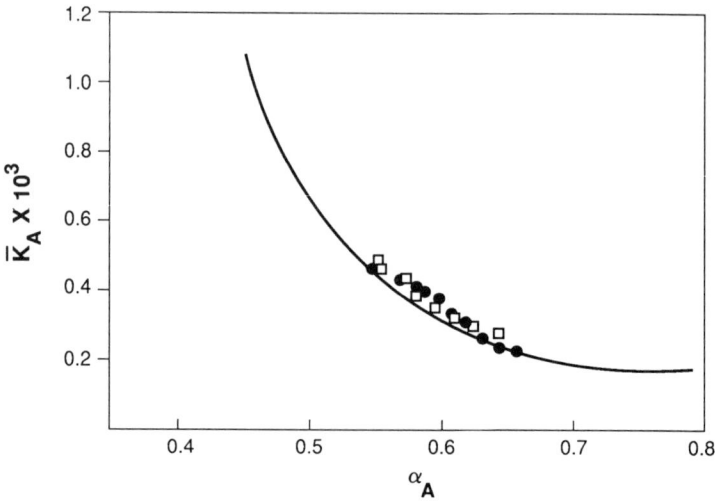

Fig. 10. Acid dissociation of fulvic acid carboxyl groups: Test of theory. Line shows predicted behavior of weighted-average equilibrium function as degree of ionization increases. Circles and squares are experimental measurements for two different fulvic acid concentrations.

2 are critically important for at least two reasons. The first is that mineral soils represent most of the usual agricultural and environmental problems. The other point is that they are the most difficult type of geochemical system to work with, so that if the research strategy works with them, then it should work for any other system.

V. Success of Research Strategy: Experimental Tests

The research strategy would not be accorded much respect if it were only an ineffectual, pedantic exercise. It will therefore be demonstrated that it has produced a continuous stream of experimental successes. These will be presented in the order of increasing complexity. This is also the order of increasing practical significance. An itemized list will present several of the examples, but the original literature will simply be cited for one or two.

A. Weighted-Average Equilibrium Functions

Burch et al. (1978) published some of the earliest proofs that exact predictive calculations can be done with a system as complicated as a fulvic acid solution. The solid curve in Fig. 10 gives the predicted behavior of the weighted-average equilibrium function for its proton dissociation. The calculations were based on previously published theory. The solid dots and squares were direct experimental measurements. The authors likewise demonstrated that dilution effects on the equilibrium can also be predicted.

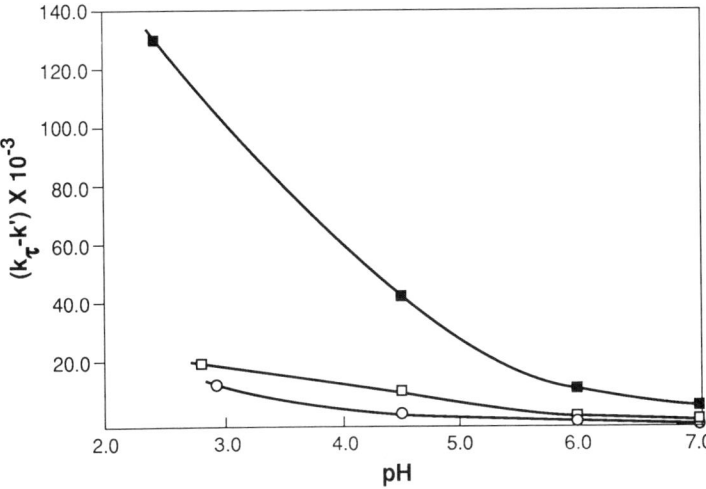

Fig. 11. Rate-constant plot for atrazine hydrolysis catalyzed by fulvic acid carboxyl groups at 25.0 °C. Use of outer variable pH produces nonlinear curves that do not penetrate the origin.

B. Chemical Equilibria in Mixed Geochemical Systems

Buffle (1988) assembled a substantial amount of data from the world literature for Cu^{++} complexing by humic materials. He found that even a partial adherence to the principle that correlated variables should directly represent cause-and-effect relationships enabled him to plot equilibrium data from many sources on the same curves. A much more complicated humic-metal ion system yielded an especially significant demonstration that classical chemical behavior can be observed in such cases. Gamble et al. (1983) developed a rigorous theoretical description of mixed equilibrium experiments in which 11 metal ions competed for the mixture of cation-exchange sites in an undissolved humic acid. The empirically tabulated experimental data failed to exhibit the classical Irving-Williams series of complexing stability constants for the metal ions. It was obscured by a combination of competing equilibria and the averaging inherent in the macroscopic measurements made on a whole mixture. A simple type of calculation based on the theory produced differential equilibrium functions that revealed the anticipated Irving-Williams series.

C. Acid Catalysis of Hydrolysis in Fulvic Acid

The acid-catalyzed hydrolysis of atrazine in fulvic acid solutions has provided a striking example of the stoichiometrically exact chemistry that can be done with complicated mixtures. Gamble and Khan (1985) found that the correct choice of variables makes it possible to simplify the graphical presentation of the rate-constant data. Figure 11 is the conventional but

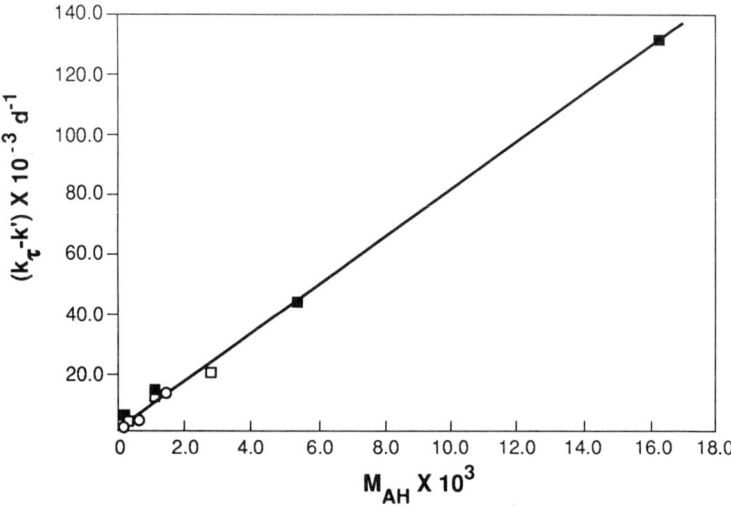

Fig. 12. Atrazine hydrolysis kinetics of Fig. 11, replotted against the inner variable, molarity of protonated carboxyl groups. Carboxyl groups are main catalyst but not properly accounted for in Fig. 11.

unrevealing plot against pH. Each of the nonlinear curves was produced by a different concentration of bulk fulvic acid. Functional group concentrations have been ignored in this conventional plot. Consequently, no curve in this graph can be used to predict any other curve. The numerical values for the rate-constant expression on the Y axis have, in fact, been caused by the protonated carboxyl groups carried by the fulvic acid and not the humic polymer itself. Thus, the rate-constant data should be plotted against the molarity of the causative carboxyl groups. Figure 12 proves this to be correct. All of the same experimental data are now on a single straight line that goes through the origin. The least-squares fitted straight line can now be used for predictive calculations. The lesson to be learned from this is that conventional thinking can sometimes be counterproductive.

D. Evidence for Generality in Mixed Geochemical Systems

Atrazine hydrolysis in systems with two physical phases has given some evidence that suggests generality. The open squares in Fig. 13 represent kinetic experiments for atrazine hydrolysis that has been catalyzed by the carboxyl groups in undissolved humic acid (Gamble and Khan 1988). A similar experiment conducted with peat soil gave the data plotted with solid circles (Gamble and Khan 1990). Within the limits of experimental error, all of the rate constants seem to have the same dependence on sorption site loading. Not only is this further evidence for the effectiveness of the research strategy, but also it should make another type of predictive calculation possible.

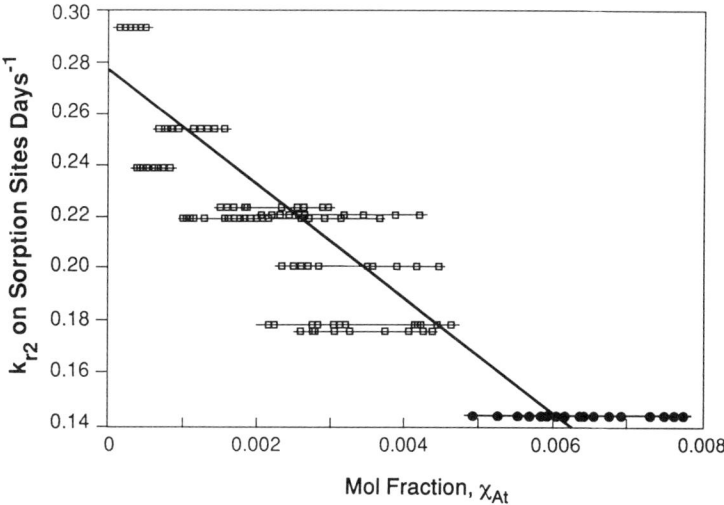

Fig. 13. Atrazine hydrolysis kinetics in systems with two physical phases. Open squares denote humic acid suspensions. Solid circles denote peat soil. χ_{At} is inner variable, mole fraction of occupied sorption sites. Its use gives a plot indicating that the two quite different samples follow the same trend.

E. Intraparticle Diffusion

For the next step in the progression to more realistic cases, theory for the intraparticle diffusion of atrazine in a mineral soil has been checked against experiment (Gamble et al. 1994). Eq. (20) has been derived from a theoretical model developed by Crank (1975)

$$\ln(\theta_L) = \ln(A) + Z \ln(t) \qquad (20)$$

In this model, steady-state labile surface sorption drives the retarded intraparticle diffusion into a semi-infinite sink. θ_L is the moles of atrazine per gram of soil taken up by intraparticle diffusion, A a constant related to the diffusion coefficient and the geometry of the surface normal to the direction of travel of the diffusing molecules, and t is time in seconds. The test of the theory is that an experimental plot of $\ln(\theta_L)$ vs. $\ln(t)$ should give a slope of $Z = \frac{1}{2}$, if the process is diffusion. A set of experiments like that in Fig. 14 has given the slopes plotted in Fig. 15. The two histogram bars on the right-hand end show a very close agreement between theory and the experimental mean. In addition, almost all of the individual experiments agreed with the theoretical value of $\frac{1}{2}$, to within the standard errors. It should be noted that the standard errors are a more demanding criterion than the standard deviations would be because it has smaller numerical values. If Crank's model applies to atrazine in a mineral soil, then the intraparticle diffusion rate should have a linear dependence on the extent

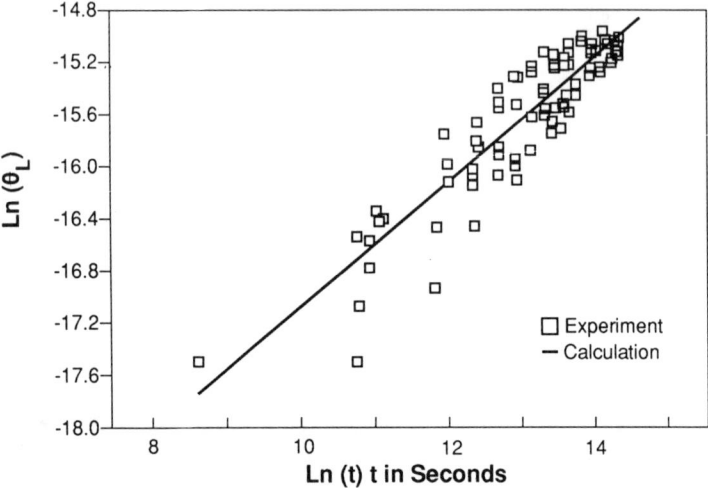

Fig. 14. Test for intraparticle diffusion, adapted from the steady-state surface sorption model of Crank (1975) (Gamble et al. 1994). See Eq. (20). An atrazine-mineral soil experiment was used. $Z = 0.49$, SE $= 0.02$. $\ln(A) = -22.0$, SE $= 0.2$.

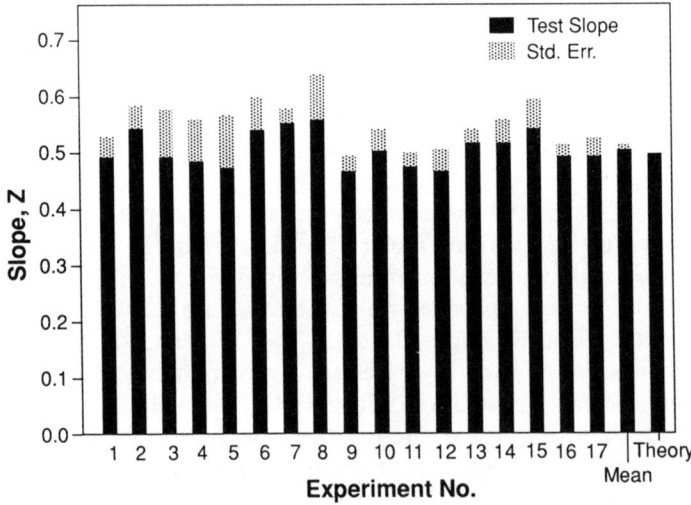

Fig. 15. Set of intraparticle diffusion tests, like that in Fig. 14 atrazine-mineral soil experiments were used.

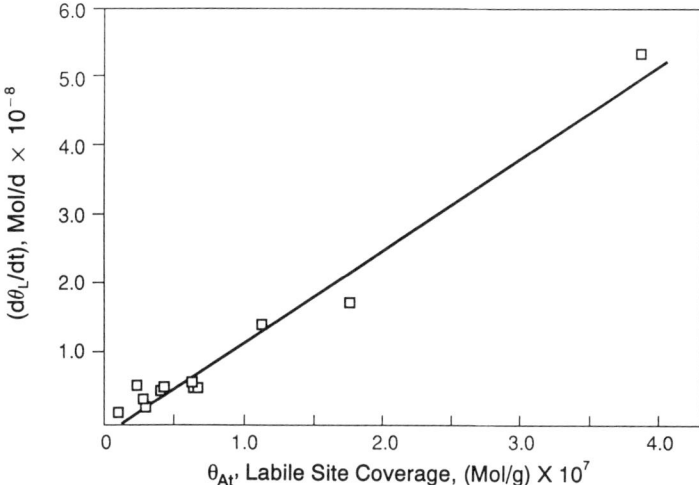

Fig. 16. Dependence of intraparticle diffusion rate on labile surface site coverage at 25.0 °C. Atrazine in a mineral soil. Diffusion rate drops to zero when there is no surface sorption on soil particles.

of surface sorption. Excellent experimental support for the model is found in Fig. 16. In this figure, θ_{At} is the labile surface sorption. The relevant point about the sorption-diffusion work is that the research strategy has now matured sufficiently so that a molecular-level mechanism might be determined.

F. Confirming Laboratory Results in the Field

The ultimate test of the research strategy is to check the laboratory research results against field experiments. Although this requires much more work, at least a few preliminary cases are available. The solid curve in Fig. 17 shows the extent to which some laboratory research currently in progress agrees with a field experiment published by Bowman (1990). In both cases, the concentrations of atrazine remaining in solution are shown. Sorption and intraparticle diffusion caused the losses. Even though this graph does not account for the effects of water as a solvent, the laboratory experiment and field measurements are not seriously divergent. The neglect of the water has been overcome by the hydrology modeling of Prasher as described by Clemente et al. (1994) in Fig. 18. The laboratory research conducted according to the concepts and methods of the research strategy has identified the mechanism for the atrazine-soil interaction. It has also calibrated the equilibrium and kinetics parameters of the mechanism. Prasher and Clemente have used this information as input for the hydrology model. Similar checks against field measurements were obtained for each of three summers. At the very least, they support the validity of the research strategy.

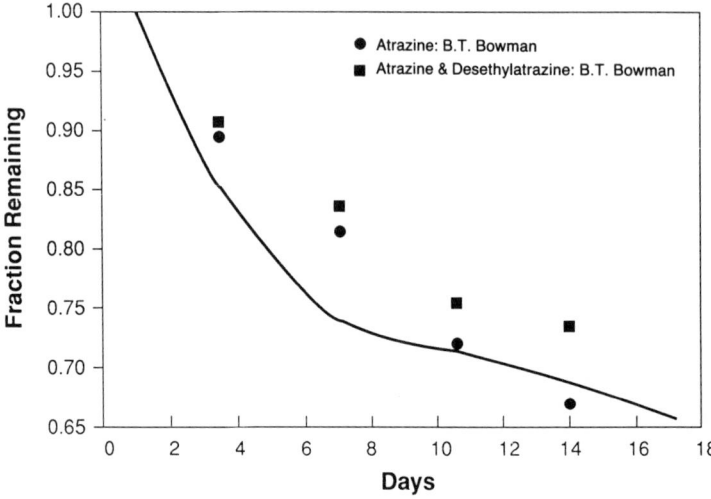

Fig. 17. Comparison of laboratory and field experiments for atrazine in mineral soils. Solid line shows results from HPLC-microfiltration method. Circles indicate field experiments showing only atrazine (Bowman 1990). Squares indicate field experiments showing atrazine + desethyl atrazine (Bowman 1990).

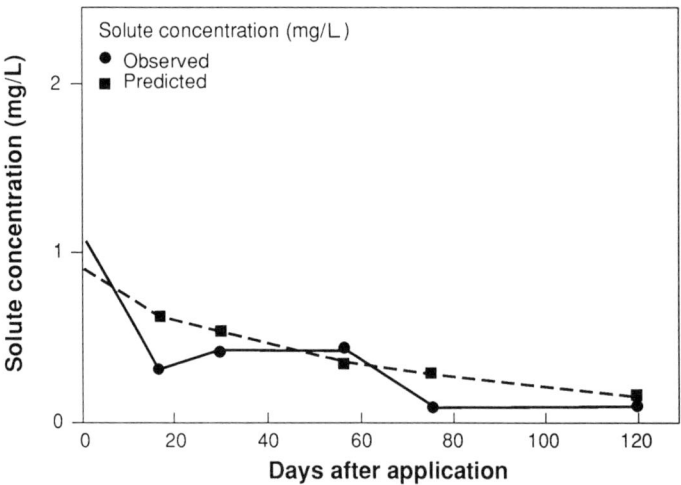

Fig. 18. Comparison of predicted and observed atrazine concentrations in the field (Gamble and Khan 1990, Gamble et al. 1983). Squares show hydrology model predictions by Prasher and Clemente using a quantitative description of the chemical mechanism developed in the laboratory. Circles indicate field measurements by Prasher and Clemente (Clemente 1991).

Summary

The purpose of environmental chemistry research should be to contribute to the solution of existing problems and prevent future problems. This will require the creation of a new multidisciplinary technology. Predictive engineering calculations are one of the means by which such a technology can be created and used. The documented successes of the research strategy imply that it will permit this to be achieved for the chemistry part of the multidisciplinary technology. An important implication is that once the concepts and methods have been set into place, the production of a much larger database will become urgent.

A new research strategy for the study of the environmental chemistry of the interaction of contaminants with soils has been developed. The purpose of this strategy is to replace chemically blind empirical work with quantitative investigations of molecular-level mechanisms. Theoretical and experimental means by which chemical stoichiometry can be applied to mixed geochemical systems are central to the strategy. An outline of the strategy is followed by some examples of the types of experimental results that can be obtained with it. Several types of experimental support for the validity and usefulness of the strategy are presented. The earliest examples included the prediction of the acid dissociation and metal ion complexing equilibria of humic materials. The kinetics of acid-catalyzed hydrolysis was next found to follow systematic trends for atrazine in soil organic matter and mineral soil. Preliminary evidence indicates that the introduction of chemical mechanisms into hydrology computer models can improve the prediction of pesticide persistence and leaching in soils. An important implication is that it is becoming possible to do predictive calculations for environmental chemistry systems. This should contribute to the solution of practical environmental problems.

References

Aiken GR, McKnight DM, Wershaw RL, MacCarthy P (eds) (1985) Humic substances in soils, sediment, and water. Wiley-Interscience, New York.

Bowman BT (1990) Mobility and persistence of alachlor, atrazine, and metolachlor in Plainfield sand, and atrazine and isazofos in Honeywood silt loam, using field lysimeters. Environ Toxicol Chem 9:453-461.

Buffle J, Altmann RJ (1987) Interpretation of metal complexation by heterogeneous complexants. In: Stumm W, (ed), Aquatic surface chemistry. Wiley-Interscience, New York, Chap 13.

Buffle J (1988) Complexation reactions in aquatic systems: An analytical approach. Ellis Horwood Ltd, Chichester, UK.

Burch RD, Langford CH, Gamble DS (1978) Methods for the comparison of fulvic acid samples: The effects of origin and concentration. Can J Chem 56:1196-1201.

Clemente R (1991) A mathematical model for simulating pesticide fate and dynamics in the environment (PESTFADE). PhD thesis, Dept of Agricultural Engineering and Faculty of Graduate Studies and Res, McGill Univ.

Clemente RS, Prasher SO, Barrington SF (1993) PESTFADE, a new pesticide fate and transport model: Model development and verification. Trans AFAE 36:357–367.

Crank J (1975) The mathematics of diffusion. Oxford Univ Press and Clarendon Press, Oxford.

de Wit JCN (1992) Proton and metal ion binding to humic substances. PhD thesis, Wageningen Agricultural Univ, Wageningen, The Netherlands.

Gamble DS (1972) Potentiometric titration of fulvic acid: Equivalence point calculations and acid functional groups. Can J Chem 50:2680–2690.

Gamble DS, Underdown AW, Langford CH (1980) Copper (II) titration of fulvic acid ligand sites with theoretical, potentiometric, and spectrophotometric analysis. Anal Chem 52:1901–1908.

Gamble DS, Schnitzer M, Kerndorff H, Langford CH (1983) Multiple metal ion exchange equilibrium with humic acid. Geochim Cosmochim Acta 47:1311–1323.

Gamble DS, Khan SU (1985) Atrazine hydrolysis in soils:Catalysis by the acidic functional groups of fulvic acid. Can J Soil Sci 65:435–443.

Gamble DS, Langford CH, Underdown AW (1985) Light scattering measurements of Cu (II)-fulvic acid complexing:The interdependence of apparent complexing capacity and aggregation. Org Geochem 8:35–39.

Gamble DS, Haniff MI, Zienius RH (1986) Solution phase complexing of atrazine by fulvic acid: A batch ultrafiltration technique. Anal Chem 54:727–731.

Gamble DS, Khan SU (1988) Atrazine hydrolysis in aqueous suspensions of humic acid at 25 °C. Can J Chem 66:2605–2617.

Gamble DS, Langford CH (1988) Complexing equilibria in mixed ligand systems: Tests of theory with computer simulations. Environ Sci Technol 22:1325–1336.

Gamble DS (1989) Titrations of organic soils with standard base: Solubility of ionizable functional groups at 25 °C. Can J Soil Sci 69:313–325.

Gamble DS, Khan SU (1990) Atrazine in organic soil: Chemical speciation during heterogeneous catalysis. J Agric Food Chem 38:297–308.

Gamble DS, Khan SU (1992) Atrazine in mineral soils: Chemical species and catalysed hydrolysis. Can J Chem 70:1597–1603.

Gamble DS, Li J, Gilchrist G, Langford CH (1994) Atrazine sorption and intraparticle diffusion in mineral soils: Mechanisms (in preparation).

Grim RE (1953) Clay mineralogy. McGraw-Hill, New York.

Harris LB (1968) Adsorption on a patchwise heterogeneous surface: Mathematical analysis of the step-function approximation to the local isotherm. Surf Sci 10:129–145.

Klotz IM, Hunston DL (1984) Mathematical models for ligand-receptor binding: Real sites, ghost sites. J Biol Chem 259:10060–10062.

Langford CH, Gutzman DW (1992) Kinetic studies of metal speciation. Anal Chim Acta 256:183–201.

Nederlof MM, van Reimsdijk WH, Koopal LK (1990) Determination of adsorption affinity distributions: A general framework for methods related to local isotherm approximations. J Colloid Interface Sci 135:410–426.

Nederlof MM (1992) Analysis of binding heterogeneity. PhD thesis, Wageningen Agricultural Univ, Wageningen, The Netherlands.

Newman ACD (1987) Chemistry of clays and clay minerals. Mineralogical Soc mono no 6, Longman Scientific & Technical, Wiley-Interscience, New York.

Pignatello J (1989) Sorption dynamics of organic compounds in soils and sediments. In: Sawhney BL, Brown K, (eds), Reactions and movement of organic chemicals in soils. SSSA special pub no 22, Soil Science Soc of America, Amer Soc of Agronomy, Madison, WI, Chap 13.

Shuman MS, Collins BJ, Fitzgerald PJ, Olson DL (1983) Distribution of stability constants and dissociation rate constants among binding sites on estuarine copper-organic complexes: Rotated disk electrode studies and an affinity spectrum analysis in selective electrode and photometric data. In: Christman RF, Gjessing ET (eds), Aquatic and terrestrial humic materials. Ann Arbor Science, Ann Arbor, MI, Chap 17.

Sojo LE (1992) Ultrafiltration as speciation tool for paraquat in humic acid suspensions: Effects of solution composition on membrane properties. Anal Chim Acta 258:219–227.

Taylor RM (1987) Non-silicate oxides and hydroxides. In: Newman ACD (ed), Chemistry of clays and clay minerals. Mineralogical Soc mono no 6, Longman Scientific & Technical, Wiley-Interscience, New York, Chap 2.

Wang Z, Gamble, DS, Cooper LH (1991) Interaction of atrazine with Laurentian humic acid. Anal Chim Acta 244:135–143.

Manuscript received August 28, 1993; accepted September 2, 1993.

Turtles as Monitors of Chemical Contaminants in the Environment

Linda Meyers-Schöne* and Barbara T. Walton†

Contents

I. Introduction 93
II. Classification of Turtles 94
   A. Freshwater Turtles 96
   B. Marine Turtles 98
   C. Terrestrial Turtles 99
III. Contaminant Residues in Turtles 99
   A. Pesticides 99
   B. Polychlorinated Biphenyls 114
   C. Dioxins and Furans 119
   D. Metals 123
   E. Radionuclides 133
IV. Other Biomonitoring Tools 142
   A. Biochemical and Histopathological Responses to Stress 142
   B. Turtle Growth Rates in Relation to Contamination 144
V. Turtles as Monitors: Advantages and Limitations 144
Summary 145
Acknowledgments 147
References 148

I. Introduction

Biota have been used increasingly in recent years to evaluate the presence of hazardous chemicals in the environment and to determine the impact of toxicants on ecosystems (e.g., Suter and Loar 1992; Hoffman et al. 1990; Clark et al. 1988; Talmage and Walton 1991; and Albers et al. 1986). Because turtles are relatively long-lived, widely distributed geographically, and found in a variety of habitats, they may be useful indicators of chemical contamination in monitoring programs. Species selection is an important consideration for a successful biological monitoring study (National Research Council 1986), and turtles offer a number of advantages as indicators of the availability of chemical and radioactive contaminants in aquatic and terrestrial ecosystems (Meyers-Schöne et al. 1993). However, not all

---

*International Technology Corporation, 5301 Central Avenue, NE, Albuquerque, NM 87108, U.S.A.
†Environmental Sciences Division, Oak Ridge National Laboratory, P.O. Box 2008, Oak Ridge, TN 37831-6038, U.S.A. Author to whom correspondence may be addressed.

© 1994 by Springer-Verlag New York, Inc.
Reviews of Environmental Contamination and Toxicology, Vol. 135.

turtle species can be expected to be equally good choices for monitoring programs because species-specific characteristics, such as differences in food habits, habitat use, home range, age, and sex, are likely to affect the exposure of individual animals at a chemically contaminated site.

The purpose of this review is to provide residue data and biological information to facilitate the selection of an appropriate turtle species for field studies designed to evaluate the bioavailability of toxicants in aquatic and terrestrial environments. An underlying premise in this review is that biological monitoring studies yield more useful data when these studies are designed with careful consideration of the characteristics of the site, properties of the chemical or radionuclide contaminant(s), and biology of the species under investigation. Not all species of a given taxon can be expected to be equally useful indicators of toxicants within a contaminated ecosystem (i.e., Talmage and Walton 1991); thus, care should be given to sample those species most likely to experience toxicant exposure as a result of what they eat and where they spend their time.

The published literature on toxicant residues in turtles and the use of biochemical analyses and growth indices as endpoints for toxicant exposure under field conditions are reviewed. General guidelines are provided for the inclusion of turtles in biological monitoring programs for chemical and radionuclide contaminants.

II. Classification of Turtles

The 250 species of turtles found worldwide all belong to the order Testudines. This order comprises species that occupy freshwater, marine, and terrestrial habitats. The amount of time turtles spend in the water varies considerably. Sea turtles, for example, venture on land only to lay their eggs in sandy beaches, whereas box turtles seldom use aquatic habitats. In addition to habitat variations among the species, turtles also differ in their food habits. Both carnivorous and herbivorous species exist; however, most species are omnivorous and ingest a variety of plant and animal matter. Diet may also change with the age of the turtle, as has been reported for the slider (*Trachemys scripta*). Juveniles are carnivorous, and adults ingest primarily vegetation (Clark and Gibbons 1969). Because habitat use is a primary determinant of turtle exposure to toxicants under natural conditions, information on the natural history of turtles is provided. This information is grouped by habitat before the discussion of body burdens of specific chemical contaminants. For convenience, freshwater turtles are subdivided into basking and bottom-dwelling species. Marine and terrestrial species are discussed separately.

Only those species found in North America are discussed. Table 1 is a listing of the scientific names and common names of the species discussed. Except where indicated otherwise, the natural history information that follows is summarized from Conant and Collins (1991).

Table 1. Turtle Nomenclature[a]

Family/Species	Common Name
Freshwater Turtles	
Emydidae	
Chrysemys picta	Painted turtle
Chrysemys picta bellii	Western painted turtle
Dierochelys reticularia	Chicken turtle
Emydoidea blandingii	Blanding's turtle
Graptemys geographica	Common map turtle
Malaclemys terrapin macrospilota	Ornate diamondback
Mauremys caspica rivulata	Caspian terrapin
Pseudemys concinna	River cooter
Pseudemys concinna suwanniensis	Suwannee river cooter
Pseudemys floridana peninsularis	Peninsula cooter
Pseudemys nelsoni	Florida redbelly turtle
Trachemys scripta	Slider
Trachemys scripta elegans	Red-eared slider
Trachemys scripta scripta	Yellowbelly slider
Chelydridae	
Chelydra serpentina	Snapping turtle
Macroclemys temminckii	Alligator snapping turtle
Kinosternidae	
Kinosternon baurii	Striped mud turtle
Kinosternon flavescens	Yellow mud turtle
Sternotherus minor minor	Loggerhead musk turtle
Sternotherus minor peltifer	Stripeneck musk turtle
Sternotherus odoratus	Common musk turtle
Kinosternon subrubrum hippocrepis	Mississippi mud turtle
Kinosternon subrubrum steindachneri	Florida mud turtle
Kinosternon subrubrum subrubrum	Eastern mud turtle
Trionychidae	
Apalone ferox	Florida softshell
Apalone mutica	Smooth softshell
Apalone spinifera	Spiny softshell
Apalone spinifera spinifera	Eastern spiny softshell
Marine Turtles	
Cheloniidae	
Caretta caretta	Loggerhead
Chelonia mydas	Green turtle
Eretmochelys imbricata	Hawksbill
Lepidochelys kempii	Atlantic ridley
Lepidochelys olivacea	Pacific ridley

(continued)

Table 1. (*Continued*)

Family/Species	Common Name
Dermochelyidae	
Dermochelys coriacea	Leatherback
Terrestrial Turtles	
Emydidae	
Terrapene carolina	Eastern box turtle
Terrapene ornata ornata	Ornate box turtle
Testudinidae	
Gopherus agassizii	Desert tortoise
Gopherus berlandieri	Texas tortoise
Gopherus polyphemus	Gopher tortoise

[a]Only those species included in this review are listed. Current scientific and common names were obtained from Collins (1990; 1993, personal communication).

A. Freshwater Turtles

Freshwater turtles consist of species that bask in the sun and those that are primarily bottom dwellers. All of the basking turtles found in North America belong to the family Emydidae. Although box turtles also belong in this family, they are discussed separately under terrestrial turtles. Bottom-dwelling species include those from the families Chelydridae, the snapping turtles; Kinosternidae, the musk and mud turtles; and Trionychidae, the softshell turtles. Freshwater turtles frequent a variety of habitats including bogs, ponds, lakes, and rivers. Regardless of the amount of time actually spent in the water, mating occurs in water, and eggs are laid in holes dug in the ground, usually near the bank. Clutch sizes are generally small, ranging from two to six eggs.

The largest of the turtle families, Emydidae, is distributed worldwide. The shells and bodies of Emydidae generally appear more colorful than those of the bottom-dwelling turtles. Species within this family occupy a large array of habitat types including marshes, bogs, ponds, lakes, rivers, and estuaries. Within this family are several genera of turtles that occur in North America, specifically *Clemmys* (spotted, bog, wood, and Western pond turtles), *Graptemys* (map turtles), *Trachemys* (sliders), *Chrysemys* (painted turtles), *Pseudemys* (cooters and redbellies), *Deirochelys* (chicken turtles), *Emydoidea* (Blanding's turtle), and *Malaclemys* (terrapins). The species most commonly used in biomonitoring studies are the sliders (*Trachemys scripta*) and the painted turtles (*Chrysemys picta*). Both species use a wide variety of habitats such as ponds, lakes, ditches, and rivers. Painted turtles and sliders tend to frequent areas where the water is somewhat shallow and aquatic vegetation is abundant. Painted turtles have a wider distribution than the sliders and occur in the midwest, in the eastern United States, and in parts of the southern and western areas of North America. Sliders are restricted to the southern region of the United States; this range extends from Virginia to Texas. Sliders tend to be slightly larger than painted turtles; adult length ranges from 12.5 to 20 cm. Adult sliders are primarily vegetarians; however, they will feed on large insects when abundant (Meyers-Schöne et al. 1993). Painted turtles are omnivorous and feed on aquatic plants and invertebrates. Both species are easily trapped with baited hoop nets.

Bottom-dwelling species live in close association with the sediment floor. They do, however, come up to the surface for air and venture on land to lay eggs and to relocate. The largest among the bottom-dwelling turtles is the alligator snapping turtle (*Macroclemys temminckii*), which may weigh from 16 to 68 kg and have body length from 38 to 66 cm. These very large, aggressive turtles are found in lakes and rivers in the United States near the southern portion of the Mississippi River. Alligator snappers are carnivores and feed primarily on fish. Also in the family Chelydridae is the snapping turtle, *Chelydra serpentina*. This aggressive species is omnivorous; however, animal matter constitutes the majority of its diet. The snapping turtle is smaller than the alligator snapping turtle and weighs from 4.5 to 16 kg as an adult. Snapping turtles occur in most states east of the Rocky Mountains and have been used to monitor a variety of chemical and radionuclide contaminants in the environment.

Three species of softshell turtles occur in North America: the smooth softshell (*Apalone mutica*), spiny softshell (*A. spinifera*), and Florida softshell (*A. ferox*). Softshell turtles can be found in rivers and lakes in the south central and southeastern United States. They have a characteristic flat, pancake-shaped shell and long, pointed snout. Females, which may exceed 35 cm in length, are often larger than males. All softshells are carnivores that consume primarily insects, crayfish, and worms (Stebbins 1985; Dalrymple 1977); however, they may also ingest some plant matter (Williams and Christiansen 1981). They have rarely been used in biological monitoring efforts.

The mud and musk turtles are relatively small bottom dwellers. Mud turtles belong to the genus *Kinosternon*, whereas musk turtles are included in the genus *Sternotherus*. Turtles from both genera are generally <10 cm in length and can be found in a variety of habitats. The most widely distributed mud turtles are members of the species *K. subrubrum*, which are found in the southern United States extending from eastern Texas to North Carolina and including regions along the eastern coast. These turtles include the Florida mud turtle (*K. s. steindachneri*), which is common to shallow bodies of water such as drainage ditches and shallow ponds; the Mississippi mud turtle (*K. s. hippocrepsi*), common in swamps; and the eastern mud turtle (*K. s. subrubrum*), abundant in tidal marshes. The Mississippi mud turtle and yellow mud turtle (*K. flavescens*) feed predominantly on arthropods, mollusks, and aquatic vegetation (Mahmoud 1968). *Kinosternon subrubrum* and the yellow mud turtle have been used in biomonitoring studies.

Musk turtles that have been used as monitors of contaminated environments are the common musk turtle (*Sternotherus odoratus*) and loggerhead musk turtle (*S. minor minor*). These turtles are frequently found in shallow, clear-water lakes, ponds, and rivers. The common musk turtle is the most widely distributed of the musk turtles and is found in portions of New England, the southeastern United States, and the states south of the Great Lakes region. The diet of the stripeneck musk turtle (*S. m. peltifer*) consists primarily of snails and insects (Folkerts 1968), whereas the common musk turtle is omnivorous (Mahmoud 1968). Both musk turtles and mud turtles are difficult to trap with hoop nets.

B. Marine Turtles

Sea turtles belong to the families Cheloniidae or Dermochelyidae. Six of the seven species of sea turtles are found in the waters of North America. All but the leatherbacks (*Dermochelys coriacea*) belong to Cheloniidae. All sea turtles have flipper-like limbs, which enable them to live their entire lives in the open seas. Sea turtles are much larger than most freshwater turtles. The Pacific ridley (*Lepidochelys olivacea*) and Atlantic ridley (*L. kempii*) weigh from 36 to 45 kg (Stebbins 1985), and the leatherbacks weigh from 295 to 544 kg (Stebbins 1985). Sea turtles mate in tropical and subtropical waters. Females return to the coastlines where they were hatched to deposit their eggs in the sandy shores. Clutch sizes for most species range from 50 to 200 eggs.

The food habits of sea turtles are species-specific. Adult green turtles (*Chelonia mydas*) are largely herbivorous, feeding preferentially on sea grass (Carr 1952). Loggerheads (*Caretta caretta*) are omnivorous; their diets consist of an assortment of crabs, mollusks, sponges, jellyfish, fish, eelgrass, and seaweed (Carr 1952; Stebbins 1985). The diets of Pacific and Atlantic ridleys are similar to that of the loggerhead. Seaweeds, shellfish, crustaceans, jellyfish, sea urchins, and fish are the primary items consumed by these species (Stebbins 1985; Dobie et al. 1961). Hawksbills (*Eretmochelys imbricata*) reside predominantly in coral reefs of tropical regions, where their diets consist primarily of sponges, including highly siliceous species (Meylan 1988). Leatherbacks, which feed almost exclusively on jellyfish, are the most pelagic of the sea turtles (Eisenberg and Frazier 1983; National Academy Press 1990).

Sea turtle eggs have been collected and used to monitor chlorinated organic compounds. In addition, considerable data are available for pesticides and heavy metals in sea turtles, especially loggerheads. Because all sea turtles found in U.S. and Central American waters are protected by law, the Department of Conservation in the state(s) of interest should be consulted on the distributions of sea turtles and applicable regulations before any sampling program is undertaken.

C. Terrestrial Turtles

Terrestrial turtles spend most of their lives on land. Unlike the aquatic species, terrestrial turtles even mate on land. Box turtles belong to the genus *Terrapene*. Subspecies exist for the two most common box turtles, *Terrapene carolina* and *T. ornata*, which occur in most states east of the Rocky Mountains. Box turtles live in a variety of habitats from wooded areas to arid grasslands. They are omnivorous; however, some species such as the ornate box turtle (*T. o. ornata*) consume primarily insects. Adult box turtles reach 10–15 cm in length and may live as long as 30–40 years. Box turtles have occasionally been used to monitor pesticide and radionuclide concentrations in the environment.

Three species of tortoises (*Testudinidae*) occur in North America. These are the gopher tortoise (*Gopherus polyphemus*), Texas tortoise (*G. berlandieri*), and desert tortoise (*G. agassizii*). The desert tortoise, which is a protected species, is the largest of the three species, with an adult size ranging from 20 to 36 cm. The gopher tortoise inhabits sandy regions of the Coastal Plains. The Texas tortoise and desert tortoise are found within limited areas in arid regions of the American Southwest. Tortoises often construct burrows, where they are concealed from the heat of the sun. All three species are herbivorous and long-lived, but they have rarely been used in biological monitoring studies.

III. Contaminant Residues in Turtles

Studies of chemical contaminants in freshwater, marine, and terrestrial turtles are presented in this review. Unfortunately, the majority of published information on contaminant concentrations in turtles does not make reference to concentrations of specific contaminants in the abiotic environment. All residue concentrations reported in the open literature since 1950 are summarized by chemical contaminant and are expressed on a wet weight basis unless noted otherwise. Included in this review are discussions of pesticides, polychlorinated biphenyls (PCBs), dioxins, furans, metals, and radionuclides in field-collected turtles.

A. Pesticides

The only pesticides that have been reported in field-collected turtles are organochlorine insecticides. A review of pesticides in reptiles was published by Hall in 1980. The present review, however, is restricted to data published on organochlorine insecticide concentrations in turtles from 1950 to 1993 as they pertain to biological monitoring. The pesticides that have been detected in turtles are 1,1'-(2,2,2-trichloroethyledene)*bis*(4-chlorobenzene)

(DDT), aldrin (1,2,3,4,10,10-hexachloro-1,4,4a,5,8,8a-hexahydro-1,4-*endo-exo*-5,8-dimethanonaphthalene), dieldrin (1,2,3,4,10,10-hexachloro-*exo*-6,7-epoxy-1,4,4a,5,6,7,8a-octahydro-1,4-*endo-exo*-5,8-dimethanonaphthalene), endrin (1,2,3,4,10,10-hexchloro-6,7-epoxy-1,4,4a,5,6,7,8,8a-octahydro-1,4-*endo-endo*-5,8-dimethanonaphthalene), heptachlor (1,4,5,6,7,8,8-heptachloro-3a,4,7,7a-tetrahydro-4,7-methano-1H-indene), mirex (1,1a,2,2,3,3a,4,5,5,-5a,5b,6-dodecachlorooctahydro-1,3,4-metheno-1H-cyclobuta(cd)pentalene), chlordane (1,2,4,5,6,7,8,8-octachloro-2,3,3a,4,7,7a-hexahydro-4,7-methano-1-1H-indene), nonachlor (1,2,3,4,5,6,7,8,8-nonachlor-2,3,3a,4,7,7a-hexahydro-4,7-methano-1H-indene), and toxaphene (a mixture of various chlorinated camphenes). Concentrations of these insecticides in turtle tissues and turtle eggs are presented in Table 2.

The most frequently monitored pesticide in turtles has been DDT and its metabolites (i.e., Meeks 1968; Hillestad et al. 1974; Reeves et al. 1977; Stone et al. 1980; Clark and Krynitsky 1980; Albers et al. 1986; Hebert et al. 1993) (Table 2). As in birds and mammals (Matsumura 1975), DDT in turtles is metabolized to dichlorodiphenyldichloroethylene (DDE) and dichlorodiphenyldichloroethane (TDE), with p,p'-DDE as the major metabolite (Owen and Wells 1976). The presence of DDT in terrestrial turtles (Stickel 1951), freshwater turtles (Meeks 1968; Flickinger and King 1972; Reeves et al. 1977; Punzo et al. 1979; Stone et al. 1980; Albers et al. 1986; Hebert et al. 1993), sea turtles (McKim and Johnson 1983), and in the eggs of marine turtles (Hillestad et al. 1974; Thompson et al. 1974; Clark and Krynitsky 1980; Clark and Krynitsky 1985) collected since the 1972 ban on the use of DDT in the United States reveals the persistence and widespread distribution of the pesticide in the environment.

In turtles, as in other vertebrate taxa, organochlorine pesticides have affinity for tissues high in lipid content (Meeks 1968; Stone et al. 1980; Pearson et al. 1973). The liver and kidneys in turtles tend to contain concentrations of these pesticides that are lower than those detected in fat, but higher than the concentrations detected in other soft tissues (Meeks 1968; Pearson et al. 1973). Concentrations of DDT and its metabolites in reproductive organs are usually within the range of concentrations detected in liver and kidney tissues (Meeks 1968). Both DDT and dieldrin have been detected in the brains of turtles exposed to these pesticides (Meeks 1968; Pearson et al. 1973). Chlorinated pesticides may eventually be eliminated in the urine and feces (Matsumura 1975) or transferred to eggs in utero (Flickinger and King 1972; Holcomb and Parker 1979; Punzo et al. 1979; Stone et al. 1980).

Differences in the concentration of an organochlorine pesticide among species of turtles collected from a common site can be related to differences in food habits. Holcomb and Parker (1979) monitored concentrations of mirex in the slider and Eastern box turtle (*Terrapene carolina carolina*) collected from an area that received four applications of the insecticide (5% mirex solution) over an 8-yr period. Higher mirex concentrations were

Table 2. Pesticide Concentrations in Field-Collected Turtles

Pesticide	Species	Location	Sex	Concentration (μg/g wet weight)	N	Observation	Reference
Aldrin	*Kinosternon flavescens*	Flooded rice field, TX	F	4.00 whole-body[a]	1	Elevated concentration	Flickinger and Mulhern (1980)
	Kinosternon flavescens	Rice field, TX	Unknown	nd,[b] whole-body	3	Not detected	Flickinger and King (1972)
	Trachemys scripta elegans[c]	Rice field, TX	Unknown	4.8 whole-body	2	Higher than in *K. flavescens*; lower in eggs than in adult *T. scripta*	Flickinger and King (1972)
				0.2 unlaid eggs[a]	4		
Chlordane	*Caretta caretta*	National Wildlife Refuge, FL	—	<0.005–0.017 eggs[a]	9	Trace amounts in 2 eggs	Clark and Krynitsky (1980)
	Caretta caretta	National Wildlife Refuge, FL	—	0.005–0.008 eggs	56	*Cis* and *oxy*-chlordane near detection limit	Clark and Krynitsky (1985)
	Chelonia mydas	National Wildlife Refuge, FL	—	<0.005 eggs	2	Not detected	Clark and Krynitsky (1980)
	Chelydra serpentina	Contaminated wetland #1 (brackish), NJ	M	1.30 fat[d]	8	*Trans*, *cis*, and *oxy* forms measured; *oxy*-chlordane predominant form with concentrations reported here; males generally higher than females	Albers et al. (1986)
			F	0.52 fat[d]	3		
		Contaminated wetland #2 (freshwater), NJ	M	0.95 fat[d]	8		
		Reference wetland, MD	M	1.27 fat[d]	7		
			F	1.09 fat[d]	6		
DDT and metabolites	*Caretta caretta*	National Wildlife Refuge, FL	—	0.047 eggs	9	DDE in all eggs; DDT in 2 eggs (nd–0.048)	Clark and Krynitsky (1980)

(*Continued*)

Table 2. (Continued)

Pesticide	Species	Location	Sex	Concentration (μg/g wet weight)	N	Observation	Reference
	Caretta caretta	National Wildlife Refuge, FL	—	0.099 eggs	56	DDE measured; did not decline during incubation	Clark and Krynitsky (1985)
	Caretta caretta	Seashores, GA, SC	—	0.058–0.305^e egg yolks	unknown	Total of DDT, DDE, and TDE reported	Hillestad et al. (1974)
	Caretta caretta	East Coast, FL	Unknown	<0.001–0.040 muscle <0.001–0.051 liver	9 8	Traces of DDE	McKim and Johnson (1983)
	Chelonia mydas	National Wildlife Refuge, FL	—	<0.005–0.042 eggs	2	DDT in 1 egg (reported here); DDE also in 1 egg at 0.005 μg/g; DDE higher in C. caretta than C. mydas	Clark and Krynitsky (1980)
	Chelonia mydas	East Coast, FL	Unknown	<0.001 muscle <0.001 liver	4 4	DDE measured but not detected at <0.001 μg/g;	McKim and Johnson (1983)
	Chelonia mydas	Ascension Island	—	nd^b–0.009 egg yolks	10	DDE in 7 eggs; DDT not detected	Thompson et al. (1974)
	Chelydra serpentina	Contaminated wetland #1 (brackish), NJ	M F	0.02 fat^d 0.06 fat^d	8 3	DDT, DDE, and DDD measured; DDE reported here was only form in all turtles; DDD at 0.1 ppm in turtles from NJ site #2	Albers et al. (1986)
		Contaminated wetland #2 (freshwater), NJ	M	1.49 fat^d	8		
		Reference wetland, MD	M F	0.25 fat^d 0.07 fat^d	7 6		
	Chelydra serpentina	Southern Ontario, Canada (16 sites)	Both	<0.01–0.165 muscle	78	Total DDT measured and range of mean concentrations at each site reported; highest individual value 1.1 μg/g	Hebert et al. (1993)

Species	Location	Sex	Concentration/Tissue	N	Comments	Reference
Chelydra serpentina	DDT radiolabeled marsh, OH	M	0.1^f brain	1	DDT highest in fat; C. serpentina contained greatest level compared with E. blandingii and C. picta	Meeks (1968)
			1.1^f liver	1		
			0.3^f kidney	1		
			0.05^f heart	1		
			13.0^f fat	1		
			0.2^f lung	1		
			2.2^g testes	1		
			0.05^f muscle	1		
			0.2^f blood	1		
			0.2^f spleen	1		
			0.2^f pancreas	1		
			0.2^f eye	1		
			0.1^f shell	1		
Chelydra serpentina	Agricultural area, IA	F	nd fat	1	DDE not detected in C. serpentina or eggs	Punzo et al. (1979)
		—	nd unlaid egg	1		
Chelydra serpentina	Tobacco fields (site #1), NC	Unknown	0.08–0.29 whole-body	2	DDT, TDE, and DDE were measured; DDE only metabolite at 0.07 ppm; only DDE reported here; no apparent site or species difference	Reeves et al. (1977)
Chelydra serpentina	Tobacco fields (site #2), NC	Unknown	0.16 whole-body	6		
Chelydra serpentina	Hudson River, NY	M	nd–57.5 fat	6	DDE highest in fat	Stone et al. (1980)
			nd–17.4 liver	11		
			nd–0.26 muscle	12		
		F	<0.05–0.99 liver	5		
			<0.05–0.74 muscle	5		
		—	<0.18 unlaid eggs	5		
Chelydra serpentina	Various waters in NY other than Hudson River	M	nd–81.3 fat	4	DDE highest in fat of male from Irondequoit Bay	Stone et al. (1980)
			nd–3.58 liver	4		
			0.023–0.042 muscle	2		
		F	4.43 fat	5		
			0.29 liver	3		
			<0.01–0.025 muscle	3		

(Continued)

Table 2. (*Continued*)

Pesticide	Species	Location	Sex	Concentration (μg/g wet weight)	N	Observation	Reference
	Chrysemys picta	DDT radio-labeled marsh, OH	M	nd brain	3	See *Chelydra serpentina*, Meeks (1968)	Meeks (1968)
				nd–0.7^f liver	4		
				nd–0.2^f kidney	4		
				nd–0.2^f heart	4		
				4.1^f fat	4		
				0.1^f lung	3		
				2.2^g testes	4		
				nd, 0.05^f muscle	4		
				0.5^f blood	1		
				nd spleen	1		
				0.2^f shell	4		
			F	nd brain	1		
				0.3^f liver	1		
				0.2^f kidney	1		
				nd heart	1		
				3.2^f fat	1		
				0.06^f lung	1		
				0.7^f ovary	1		
				0.05^f muscle	1		
				0.2^f blood	1		
				0.1^f spleen	1		
				nd pancreas	1		
				0.3^f eye	1		
				0.09^f shell	1		
			Unknown	nd brain	2		
				0.1–0.2^f liver	2		
				nd kidney	2		
				nd heart	2		

Species	Location	Sex	Concentration	Notes	n	Reference
			0.5–1.5^f fat		2	
			nd–0.1^f lung		2	
			nd muscle		2	
			nd–0.06^f blood		2	
			nd spleen		2	
			nd pancreas		1	
			nd shell		2	
Chrysemys picta	Lake, TN	Unknown	0.15 liver	DDE was predominant form; highest in fat; mean concentrations higher in *T. scripta* than *C. picta*	2	Owen and Wells (1976)
			nd brain		2	
			2.67 fat		2	
			nd–0.004 excrement		2	
Chrysemys picta	Agricultural area, IA	M	0.018 fat	Trace of DDE	1	Punzo et al. (1979)
Chrysemys picta	Tobacco fields (site #1), NC	Unknown	0.43 whole-body	See *Chelydra serpentina*, Reeves et al. (1977)	3	Reeves et al. (1977)
	Tobacco fields (site #2), NC	Unknown	0.64 whole-body		8	
Emydiodea blandingii^h	DDT radiolabeled marsh, OH	M	nd brain	See *Chelydra serpentina*, Meeks (1968)	1	Meeks (1968)
			0.9^f liver		1	
			0.4^f kidney		1	
			0.09^f heart		1	
			3.8^f fat		1	
			0.08^f lungs		1	
			1.5^g testes		1	
			0.03^f muscle		1	
			0.2^f blood		1	
			nd spleen		1	
			0.1^f pancreas		1	
			0.09^f shell		1	

(Continued)

Table 2. (Continued)

Pesticide	Species	Location	Sex	Concentration (μg/g wet weight)	N	Observation	Reference
			F	0.3^f brain	1		
				0.9^f liver	1		
				0.2^f kidney	1		
				0.1^f heart	1		
				4.6^f fat	1		
				0.2^f lung	1		
				0.05^f muscle	1		
				0.5^f ovary	1		
				0.6^f shell	1		
	Kinosternon flavescens	Rice field, TX	Unknown	1.2 whole-body	3	Higher in *K. flavescens* than *T. scripta*	Flickinger and King (1972)
	Sternotherus odoratus	Tobacco fields (site #1), NC	Unknown	0.36 whole-body	4	See *Chelydra serpentina*, Reeves et al. (1977)	Reeves et al. (1977)
		Tobacco fields (site #2), NC	Unknown	0.89 whole-body	6		
	Terrapene carolina carolina	Forest area treated with DDT, MD	Both	Not measured	86	No difference in population size; growth of 4 juveniles from DDT area appeared normal	Stickel (1951)
		Reference forest, MD	Both	Not measured	82		
	Trachemys scripta^i	Lake, TN	Unknown	0.39 liver	3	DDE only form detected; highest in fat	Owen and Wells (1976)
				nd–0.33 brain	3		
				4.56 fat	3		
				nd excrement	3		
	Trachemys scripta^j	Tobacco fields (site #1), NC	Unknown	0.12 whole-body	7	See *Chelydra serpentina*, Reeves et al. (1977)	Reeves et al. (1977)
		Tobacco fields (site #2), NC	Unknown	0.71 whole-body	4		
	Trachemys scripta elegans^c	Rice field, TX	Unknown	<0.1 whole-body	2	Higher in the eggs	Flickinger and King (1972)
			—	0.2 unlaid eggs	4		

Compound	Species	Location	Sex	Concentration	N	Comments	Reference
Dieldrin	*Caretta caretta*	National Wildlife Refuge, FL	—	<0.005–0.028 eggs	9	Trace amounts in 4 eggs	Clark and Krynitsky (1980)
	Caretta caretta	Seashores, GA, SC	—	Trace–0.0564^e egg yolks	Unknown	Trace amounts	Hillestad et al. (1974)
	Chelonia mydas	National Wildlife Refuge, FL	—	<0.005 eggs	2	Not detected	Clark and Krynitsky (1980)
	Chelydra serpentina	Contaminated wetland #1 (brackish), NJ	M	nd fat^d	8	Trace amounts in males from freshwater NJ site	Albers et al. (1986)
			F	nd fat^d	3		
		Contaminated wetland #2 (freshwater), NJ	M	0.05 fat^d	8		
	Chelydra serpentina	Reference wetland, MD	M	0.02 fat^d	7	Trace amounts in males only	Albers et al. (1986)
			F	nd fat^d	6		
	Chelydra serpentina	Agricultural area, IA	F	nd fat	1	Of reptiles and amphibians examined, highest in snakes	Punzo et al. (1979)
			—	nd unlaid egg	1		
	Chelydra serpentina	Tobacco fields (site #1), NC	Unknown	<0.01 whole-body	2	Not detected	Reeves et al. (1977)
		Tobacco fields (site #2), NC	Unknown	<0.01 whole-body	6	Not detected	
	Chelydra serpentina	Hudson River, NY	M	8.41 fat	6	Highest in fat of both sexes	Stone et al. (1980)
				nd–0.16 liver	5		
				nd–0.034 muscle	6		
			F	17.0 fat	1		
				nd–0.026 liver	3		
				nd muscle	2		
			—	<0.035 unlaid eggs	6		
	Chelydra serpentina	Various waters in NY other than the Hudson River	M	nd–34.1 fat	4	Highest in fat of a male from Irondequoit Bay (Lake Ontario)	Stone et al. (1980)
				nd–0.99 liver	4		
				nd–0.16 muscle	2		
			F	nd–2.40 fat	5		
				0.06 liver	3		
				<0.01–0.01 muscle	3		

(*Continued*)

Table 2. (*Continued*)

Pesticide	Species	Location	Sex	Concentration (μg/g wet weight)	N	Observation	Reference
	Chrysemys picta bellii[j]	Agricultural area, IA	M	0.074 fat	1	See *Chelydra serpentina*, Punzo et al. (1979)	Punzo et al. (1979)
	Chrysemys picta	Tobacco fields (site #1), NC	Unknown	<0.01 whole-body	3	Not detected	Reeves et al. (1977)
		Tobacco fields (site #2), NC	Unknown	<0.01 whole-body	8	Not detected	
	Kinosternon flavescens	Flooded rice field, TX	F	47 whole-body	1	Epoxidation of aldrin to dieldrin may be a slow process in turtles	Flickinger and Mulhern (1980)
	Kinosternon flavescens	Rice field, TX	Unknown	0.6 whole-body	3	Higher in *T. scripta*	Flickinger and King (1972)
	Sternotherus odoratus	Tobacco fields (site #1), NC	Unknown	<0.01 whole-body	4	Not detected	Reeves et al. (1977)
		Tobacco fields (site #2), NC	Unknown	<0.01 whole-body	6	Not detected	
	Trachemys scripta[i]	Tobacco fields (site #1), NC	Unknown	<0.01 whole-body	8	Not detected	Reeves et al. (1977)
		Tobacco fields (site #1), NC	Unknown	<0.01 whole-body	4	Not detected	
		Tobacco fields (site #2), NC	Unknown	<0.01 whole-body	4	Not detected	
	Trachemys scripta elegans[c]	Rice field, TX	Unknown —	1.2 whole-body 2.8 unlaid eggs	2 4	Higher in eggs than adults	Flickinger and King (1972)
Endrin	*Chelydra serpentina*	Contaminated wetland, (brackish) #1, NJ Contaminated wetland, (freshwater) #2, NJ	M F M	nd fat[d] nd fat[d] nd fat[d]	8 3 8	Not detected in turtles from either site	Albers et al. (1986)
	Chelydra serpentina	Reference wetland, MD	M F	nd fat[d] nd fat[d]	7 6	Not detected	Albers et al. (1986)

Turtles as Environmental Monitors 109

	Chelydra serpentina	Tobacco fields (site #1), NC	Unknown	<0.01 whole-body	2	Not detected	Reeves et al. (1977)
		Tobacco fields (site #2), NC	Unknown	<0.01 whole-body	6	Not detected	
	Chrysemys picta	Tobacco fields (site #1), NC	Unknown	<0.01 whole-body	3	Not detected	Reeves et al. (1977)
		Tobacco fields (site #2), NC	Unknown	<0.01 whole-body	8	Not detected	
	Kinosternon flavescens	Flooded rice field, TX	F	1.3 whole-body	1	Above background	Flickinger and Mulhern (1980)
	Sternotherus odoratus	Tobacco fields (site #1), NC	Unknown	<0.01 whole-body	4	Not detected	Reeves et al. (1977)
		Tobacco fields (site #2), NC	Unknown	<0.01 whole-body	6	Not detected	
	Trachemys scripta[i]	Tobacco fields (site #1), NC	Unknown	<0.01 whole-body	8	Not detected	Reeves et al. (1977)
		Tobacco fields (site #2), NC	Unknown	<0.01 whole-body	4	Not detected	
Heptachlor (Heptachlor epoxide)	*Caretta caretta*	National Wildlife Refuge, FL	—	<0.005–0.006 eggs	9	In only 2 eggs	Clark and Krynitsky (1980)
	Chelonia mydas	National Wildlife Refuge, FL	—	<0.005	2	Not detected	Clark and Krynitsky (1980)
	Chelydra serpentina	Contaminated wetland #1 (brackish), NJ	M	nd fat[d]	8	Contaminant in males from freshwater site	Albers et al. (1986)
			F	nd fat[d]	3		
		Contaminated wetland #2 (freshwater), NJ	M	0.28 fat[d]	8		
	Chelydra serpentina	Reference wetland, MD	M	0.11 fat[d]	7	Trace in both sexes	Albers et al. (1986)
			F	0.03 fat[d]	6		

(*Continued*)

Table 2. (*Continued*)

Pesticide	Species	Location	Sex	Concentration (μg/g wet weight)	N	Observation	Reference
	Chelydra serpentina	Land treated with heptachlor, CO	Unknown	Not measured	1	Dead after 2 treatments at 0.28 kg/ha	Ferguson (1963)
	Chelydra serpentina	Agricultural area, IA	F	nd fat nd unlaid eggs	1	Heptachlor epoxide not detected	Punzo et al. (1979)
	Chrysemys picta bellii[j]	Agricultural area, IA	M	nd fat	1	Higher in snakes than turtles	Punzo et al. (1979)
	Terrapene carolina	Land treated with heptachlor, CO	Unknown	Not measured	1	Dead after 2.24 kg/ha land treatment	Ferguson (1963)
	Trachemys scripta elegans[k]	Land treated with heptachlor, GA and AL	Unknown	172.0 whole-body	1	Dead 279 d after 2.24 kg/ha application; heptachlor epoxide measured; highest among mammals and reptiles	
Mirex	*Caretta caretta*	National Wildlife Refuge, FL	—	<0.005–0.005 eggs	9	In only 1 egg	Clark and Krynitsky (1980)
	Chelonia mydas	National Wildlife Refuge, FL	—	<0.005 eggs	2	Not detected	Clark and Krynitsky (1980)
	Chelydra serpentina	Southern Ontario, Canada (16 sites)	Both	<0.1–0.004 muscle	78	Range of means at each site reported; only traces detected; highest measured was 0.009 μg/g	Hebert et al. (1993)

Terrapene carolina	Land treated with mirex over a period of 8 yr, MI	Unknown	1.1 liver	6	Higher in *T. carolina* livers than in those of *T. scripta*; also true for eggs	Holcomb and Parker (1979)
		Unknown	1.7 liver	3		
		Unknown	1.3 liver	1		
		Unknown	0.29 liver	5		
		—	1.4^g unlaid eggs	Unknown		
		—	1.6^g unlaid eggs	Unknown		
		—	2.5^g unlaid eggs	Unknown		
		—	1.4^g unlaid eggs	Unknown		
Trachemys scripta^i	Land treated with mirex over a period of 8 yr, MI	Unknown	0.41 liver	2	Peak levels in livers of both species 2.5 yr after last treatment; higher in *Terrapene carolina* than *Trachemys scripta*	Holcomb and Parker (1979)
		Unknown	0.88 liver	3		
		Unknown	0.16 liver	4		
		Unknown	0.05 liver	9		
		Unknown	0.004 liver	6		
		—	1.8^g unlaid eggs	Unknown		
		—	2.2^g unlaid eggs	Unknown		
		—	0.15^g unlaid eggs	Unknown		
		—	0.16^g unlaid eggs	Unknown		
		—	0.04^g unlaid eggs	Unknown		

Nonachlor

Caretta caretta	National Wildlife Refuge, FL	—	<0.005–0.009 eggs	9	In only 1 egg	Clark and Krynitsky (1980)
Chelonia mydas	National Wildlife Refuge, FL	—	<0.005	2	Not detected	Clark and Krynitsky (1980)
Chelydra serpentina	Contaminated wetland #1 (brackish), NJ	M	1.05 fat^d	8	*Trans* and *cis* forms; *Trans* isomer slightly higher; total of the 2 isomers reported here; males higher than females	Albers et al. (1986)
	Contaminated wetland #1 (brackish), NJ	F	0.39 fat^d	3		
	Contaminated wetland #2 (freshwater), NJ	M	0.73 fat^d	8		
	Reference wetland, MD	M	0.92 fat^d	7		
	Reference wetland, MD	F	0.46 fat^d	6		

(Continued)

Table 2. (Continued)

Pesticide	Species	Location	Sex	Concentration (μg/g wet weight)	N	Observation	Reference
Toxaphene	*Chelydra serpentina*	Contaminated wetland #1 (brackish), NJ	M	nd fat[c]	8	Not detected in animals from contaminated and reference sites	Albers et al. (1986)
			F	nd fat[c]	3		
		Contaminated wetland #2 (freshwater), NJ	M	nd fat[c]	8		
		Reference wetland MD	M	nd fat[c]	7		
			F	nd fat[c]	6		
	Kinosternon flavescens	Flooded rice field, TX	F	0.3 whole-body	1	Toxaphene present	Flickinger and Mulheren (1980)
	Kinosternon flavescens	Rice field, TX	Unknown	nd whole-body	3	Not detected	Flickinger and King (1972)
	Trachemys scripta elegans[c]	Rice field, TX	Unknown	nd whole-body	2	Not detected	Flickinger and King (1972)
			—	nd unlaid eggs	4		

[a] Shell removed or assumed to be removed prior to analysis in all cases.
[b] nd = Not detectable. Detection limit not reported.
[c] Species listed as *Pseudemys scripta elegans* in reference.
[d] Concentrations in fat estimated based on ppm of lipid concentrations and percentage of lipid in fat values.
[e] Weight basis (dry or wet) not reported.
[f] Dry weight concentrations converted to wet weight using conversion factors listed in Meeks (1968).
[g] Concentration based on dry weight.
[h] Species listed as *Emys blandingii* in reference.
[i] Species listed as *Chrysemys scripta* in reference.
[j] Species listed as *Chrysemys picta belli* in reference.
[k] Species referred to as red-eared turtle in reference.

detected in the more carnivorous box turtle than in the more vegetarian pond slider. Differences were also reported in the concentrations of DDE in the eggs of marine turtles (Clark and Krynitsky 1980). The eggs of the loggerhead sea turtle contained higher concentrations of DDE than those of the green sea turtle collected from the same seashores. Loggerheads feed on fish, clams, sponges, jellyfish, and vegetation; green sea turtles feed primarily on large seaweed (Carr 1952).

Flickinger and King (1972) also attributed a species difference in aldrin and dieldrin concentrations to the food habits of the turtles examined. Higher concentrations of these pesticides in sliders than in yellow mud turtles were attributed to the consumption of aldrin-treated rice seed by the herbivorous slider. These data are consistent with the interpretation that dietary preferences influence contaminant concentration in turtles.

Freshwater turtles have been shown to accumulate organochlorine insecticides, specifically DDT, from a contaminated wetland. Meeks (1968) treated a 1.61-ha marsh with 0.22 kg of DDT/ha and reported that during a 15-mon study the highest concentrations detected among the reptiles, birds, and mammals sampled occurred in fat from a Northern water snake (*Natrix sipedon sipedon*, 23.66 µg/g), a Virginia rail (*Rallus limicola*, 15.96 µg/g), and a snapping turtle (13.04 µg/g). Two Blanding's turtles (*Emydoidea blandingii*) and seven painted turtles contained concentrations lower than that detected in the snapping turtle. Concentrations detected in fish were generally lower than those reported in birds and reptiles. Carnivorous species such as the northern water snake, Virginia rail, and snapping turtle reached their peak concentrations 13-15 mon following initial exposure to the DDT. The data in this study are consistent with trophic levels of the turtle species influencing DDT residues in the turtles.

Interspecific differences have also been noted in the sensitivities of turtles to organochlorine pesticides. Phillips and Wells (1974) found species differences in the response of adenosine triphosphatase (ATPase) to DDT. Of the five species studied, the snapping turtle and eastern spiny softshell turtle (*Apalone spinifera spinifera*) appeared to be the most sensitive species displaying the greatest inhibition of total ATPase activity. A similar study was conducted by Wells et al. (1974), in which the effect of dieldrin and aldrin on ATPase activity in tissues of the common map turtle (*Graptemys geographica*) was measured. The degree of inhibition, however, was not as great as that observed using DDT in the map turtle (Phillips and Wells 1974), especially for kidney and liver tissue.

The data from the reviewed papers indicate several trends with regard to chlorinated organic insecticides in turtles. As has been observed in mammals and birds, organochlorines are concentrated in the adipose tissues of turtles. Such compounds can also be transferred from a gravid female to her eggs. The elimination of organochlorines into eggs may explain the higher concentrations observed in male versus female turtles of the same species from a given location. Finally, differences in organochlorine concentrations have been reported between species. These differences are often related to differences in the food habits of turtles.

B. Polychlorinated Biphenyls

Turtles appear to be excellent sentinels for PCB contamination in aquatic environments (i.e., Stone et al. 1980; Albers et al. 1986). Concentrations of PCB measured in turtle tissues are reported in Table 3. PCBs concentrate primarily in adipose tissue (Stone et al. 1980; Helwig and Hora 1983; Watson et al. 1985; Bryan et al. 1987a). Reproductive organs in turtles from PCB-contaminated sites also contain relatively high concentrations of PCBs compared with other soft tissues (Bryan et al. 1987a). The highest reported concentration of PCBs in turtles is 4530 μg/g in the fat of a snapping turtle collected from a pond near a liquid waste disposal site (Watson et al. 1985). The concentration in the fat of this turtle was 10^4 times greater than that in whirligig beetles (Gyrinidae), 4.6×10^3 to 8.7×10^4 times that measured in frogs and tadpoles (*Rana clamitans melanota* and *R. catesbeiana*), and 1.7×10^4 times that detected in three species of fish (the pumpkinseed sunfish, *Lepomis gibbosus*; golden shiner, *Notemigonius crysoleucas*; and brown bullhead, *Ictalurus nebulosus*), the only other organisms collected from this pond. The PCB concentration in turtle muscle was lower than that detected in turtle fat and was 6.3 times greater than the average PCB concentration in whole fish. Several other studies have reported PCB concentrations that exceed 1000 μg/g in the fat of field-collected turtles (Olafsson et al. 1983; Stone et al. 1980; Bryan et al. 1987a). In comparison with other vertebrates, fish are the only other taxa in which PCB concentrations exceeding 1000 μg/g have been detected in species from the wild (Eisler 1986c). These fish were collected from the Hudson River in an area of known PCB contamination (Brown et al. 1985). The ability of turtles to store high concentrations of PCBs in their fat without apparent adverse effects makes this taxon extremely useful for the biological monitoring of PCBs in freshwater ecosystems.

PCBs can be transferred from gravid turtles to eggs in utero (Stone et al. 1980; Hebert et al. 1993). Highly significant correlations have been reported for the concentrations of total PCBs in adult female muscle tissue and the eggs within these turtles (Hebert et al. 1993). PCB concentrations in the eggs appear to depend on whether fat reserves are present in the gravid females. Stone et al. (1980) reported PCB concentrations in the eggs and tissues of five gravid snapping turtles collected from the Hudson River. Liver and muscle were analyzed from each of these turtles, but a fat sample was analyzed from only one turtle. It should be noted that female turtles fast during the nesting season, which may result in the mobilization of the stored PCBs and alteration of the ratio of PCB concentrations in different tissues. Muscle contained 60.8% of the PCB concentration detected in the

Table 3. Polychlorinated Biphenyl Concentrations in Field-Collected Turtles

PCB	Species	Location	Sex	Concentration (μg/g wet weight)	N	Observation	Reference
Aroclors 1248 and 1254	*Chelonia mydas*	Ascension Island	—	0.008 egg yolks	10	Aroclor 1242 and 1254 major PCBs; mean concentration of both reported	Thompson et al. (1974)
Aroclors 1260 and 1254	*Chelydra serpentina*	5 rivers and 1 lake site, MN	M	<0.025–0.073 muscle 19.9 fat	10 9	Higher in fat than muscle	Helwig and Hora (1983)
			F	<0.025–0.086 muscle 31.2 fat	5 5		
			Unknown	<0.025 muscle <0.2 fat	2 1		
Aroclor 1260	*Caretta caretta*	National Wildlife Refuge, FL	—	0.078 eggs[a]	9	Ranged from 0.032–0.201 μg/g, higher than in *Chelonia mydas*	Clark and Krynitsky (1980)
Aroclor 1260	*Chelonia mydas*	National Wildlife Refuge, FL	—	<0.025	2	Not detected	Clark and Krynitsky (1980)
Total	*Caretta caretta*	East Coast, FL	Unknown	0.008 muscle <0.005–0.133 liver	9 8	Higher in liver than muscle	McKim and Johnson (1983)
Total	*Chelonia mydas*	East Coast, FL	Unknown	0.0018 muscle <0.005–0.070 liver	4 4	Slightly higher PCB in *C. caretta*	McKim and Johnson (1983)

(*Continued*)

Table 3. (*Continued*)

PCB	Species	Location	Sex	Concentration (μg/g wet weight)	N	Observation	Reference
Total	*Chelydra serpentina*	Contaminated wetland #1 (brackish), NJ	M F	40.41 fat[b] 8.41 fat[b]	8 3	Males from brackish-water NJ site (1) had highest concentrations and significantly higher than in males from freshwater MN site (2) and MD site	Albers et al. (1986)
		Contaminated wetland #2 (freshwater), NJ	M	17.28 fat[b]	8		
		Reference wetland, MD	M F	26.08 fat[b] 25.68 fat	7 6		
Total	*Chelydra serpentina*	Contaminated dump site, NY	M	1600 fat 100 testes 82 brain 72 liver 49 heart 48 kidney 48 pancreas 13 lungs	1 1 1 1 1 1 1 1	Highest in fat > testes > brain > liver	Bryan et al. (1987a)
Total	*Chelydra serpentina*	Reference pond, NY	M	4.2 fat 1.6 testes 1.0 brain 1.0 liver 0.64 heart 1.2 kidney 1.2 pancreas 0.41 lungs	1 1 1 1 1 1 1 1	Very low concentrations	Bryan et al. (1987a)

Total	*Chelydra serpentina*	Upper Hudson River, NY	—	1.11–2.86 egg yolks 0.12–0.48 egg whites and shell	2 2	Values are total of 5 isomers; 95% of total toxicity to egg (based on toxic equivalent of TCDD) in yolk	Bryan et al. (1987b)
Total	*Chelydra serpentina*	Southern Ontario, Canada (16 sites)	Both	0.007–0.655 muscle	78	Range of means reported; highest individual total PCB 2.12 $\mu g/g$; highly significant correlations between PCB in adult females and their eggs, and between muscle and eggs; muscle was also correlated with concentrations in liver; several PCBs significantly correlated with age, length, and weight	Herbert et al. (1993)
Total	*Chelydra serpentina*	Upper Hudson River, NY Lake Ontario	M M	3608 fat 633 fat	1 1	Elevated concentrations from both sites	Olafsson et al. (1983)
Total	*Chelydra serpentina*	Hudson River, NY	Unknown	750 fat	1	PCBs and polychlorinated dibenzofurans in fat	Rappe et al. (1981)

(*Continued*)

Table 3. (Continued)

PCB	Species	Location	Sex	Concentration (µg/g wet weight)	N	Observation	Reference
Total	Chelydra serpentina	Hudson River, NY	M	3560 fat	8	Highest in fat	Stone et al. (1980)
				82.2 liver	14		
				3.3 muscle	15		
			F	1123 fat	1		
				38.5 liver	7		
				5.4 muscle	6		
			—	28.9 unlaid eggs	6		
Total	Chelydra serpentina	Various waters in NY other than the Hudson River	M	745 fat	4	Highest in fat; concentrations in muscle within FDA limit for edible fish	Stone et al. (1980)
			F	7.4 liver	4		
				0.36–0.48 muscle	2		
				240 fat	5		
				8.1 liver	4		
				0.46 muscle	4		
Total	Chelydra serpentina	Ponds near industrial chemical disposal area, NY	Unknown	4530 fat	1	Turtle containing 81 µg/g PCBs in fat also contained 5.65 µg/g nonortho PCBs; Biomagnification reported	Watson et al. (1985)
				185 liver	1		
				17 muscle	1		
				81 fat	1		
Total	Mouremys caspica rivulata	Sewage canal of Kibutz Mishmar Hasharon, Israel	Unknown	23.2 liver	10	Mean total PCB reported; these served as controls in enzyme and cytochrome P-450 induction study	Yawetz et al. (1983)

[a]Shell was removed or assumed to be removed prior to analysis in all cases.
[b]Concentrations in fat estimated based on ppm of lipid concentrations and percentage of lipid in fat values.

liver of the turtle that had fat, and the concentration in the eggs slightly exceeded that detected in muscle. The eggs from this turtle contained 2.9% of the concentration detected in the adipose tissue of the female. In contrast, the remaining four turtles contained muscle concentrations of 10.5% of that detected in liver; PCB concentrations in eggs exceeded those reported in muscle by 10-64 times. It may be that the use of fat reserves resulted in the mobilization and preferential deposition of PCBs into the lipid-rich egg yolks. Thus, PCBs are transferred to the eggs, the majority (95%) of which is concentrated in yolk (5% is partitioned into albumin and shell) (Bryan et al. 1987b). Because high concentrations of PCBs can be transferred to turtle eggs in utero, studies are needed to determine the concentrations of PCBs that impair turtle development in the eggs and that may subsequently result in population declines in highly contaminated areas.

There is an indication that gender influences PCB accumulation in turtles. Albers et al. (1986) reported that male snapping turtles contained significantly higher concentrations of PCBs than females. The mean PCB concentration in the fat from snapping turtles was 40.4 $\mu g/g$ in males and 8.41 $\mu g/g$ in females. This difference may be due to the elimination of PCBs into eggs by mature females. If a gender difference does exist and is due to elimination of PCBs into eggs, the sampling and analysis of exclusively adult males may result in a more uniform set of data among the turtles sampled throughout the year. In any event, the gender of all turtles should be recorded when they are sampled for environmental monitoring.

Parameters other than gender have also been correlated with residue concentrations in turtles. In a study by Hebert et al. (1993), the concentrations of several PCB congeners in muscle tissues from 78 snapping turtles were found to be significantly correlated with turtle age, length, and weight (Table 3).

The literature reveals turtles as excellent monitors of PCBs. Individuals have been found containing >1000 $\mu g/g$ of PCBs in fat tissue without apparent adverse effects. PCBs are concentrated in adipose tissue. As was observed with chlorinated organic insecticides, male turtles often contain higher concentrations of PCBs than do females of the same species. Gender differences have been attributed to elimination of PCBs by gravid females to their eggs. Age and size of an individual turtle have also been positively correlated with the concentration of PCBs in muscle tissues.

C. Dioxins and Furans

Polychlorinated dibenzodioxins (PCDDs) and polychlorinated dibenzofurans (PCDFs) have occasionally been measured in turtles (Olie et al. 1989; Ryan et al. 1986; Rappe et al. 1981; Watson et al. 1985). Residue concentrations of PCDDs and PCDFs in turtles tissues are presented in Table 4. These limited data suggest tissue distributions of PCDDs and PCDFs similar to those reported for other chlorinated organics (Ryan et al. 1986).

Table 4. Dioxin and Furan Concentrations in Field-Collected Turtles

Compound	Species	Location	Sex	Concentration (pg/g wet weight)	N	Observation	Reference
Total PCDD	Unknown	South Vietnam	F	5.24[a] fat 8.82[a] muscle 26.49[a] liver 11.41[a] gall bladder 122.85[a] ovaries	1 1 1 1 1	Major PCDDs; 2,3,7,8-tetraCDD, 1,2,3,6,7,8-hexaCDD, and 1,2,3,7,8,9-hexaCDD; TCDD toxic equivalents highest in ovaries	Olie et al. (1989)
Total PCDD	*Chelydra serpentina*	St. Lawrence River, NY, site 1	M	596.3 fat 153.5 liver	1 1	Tetra and octa-PCDDs measured; in each case 2,3,7,8-tetraCDD was predominant; Distribution of congeners was similar to that in fish; Normal in appearance and behavior; bioaccumulation of TCDD suggested	Ryan et al. (1986)
		St. Lawrence River, NY, site 2	F	334 fat 47 liver	1 1		
		St. Lawrence River NY, site 3	F	596.7 fat 117.7 liver	1 1		
	Emydoidea blandingii[b]	Ottawa River, Canada	M	16 liver	1		
Total PCDF	*Chelydra serpentina*	Hudson River, NY	Unknown	3000 fat	1	Major PCDFs; 2,3,7,8-tetraCDF at 45 pg/g and 2,3,4,7,8-pentaCDF at 620 pg/g; concentrations in fat were correlated with PCBs in environment	Rappe et al. (1981)

Compound	Species	Location	Sex	Concentration	n	Comments	Reference
Total PCDF	*Chelydra serpentina*	St. Lawrence River, NY, site 1	M	164 fat 29 liver	1 1	Tetra and octa-PCDFs were measured; in all, turtles from the St. Lawrence River 2,3,4,7,8-pentaCDF was predominant; Normal in appearance and behavior	Ryan et al. (1986)
		St. Lawrence River, NY, site 2	F	101 fat 13 liver	1 1		
		St. Lawrence River, NY, site 3	F	4909 fat 827 liver	1 1		
Total PCDF	*Emydoidea blandingii*[b]	Ottawa River, Canada	M	<2 liver	1	Not detected	Ryan et al. (1986)
Total PCDF	Unknown	South Vietnam	F	3.27[a] fat 2.57[a] muscle 13.72[a] liver 19.05[a] gall bladder 100.18[a] ovaries	1 1 1 1 1	Major PCDFs: were 2,3,7,8-tetraCDF, 1,2,3,7,8-pentaCDF plus 1,2,3,4,8-pentaCDF, and 2,3,4,7,8-pentaCDF; TCDD toxic equivalents highest in ovaries	Olie et al. (1989)
2,3,7,8-TCDF	*Chelydra serpentina*	Pond near industrial chemical disposal area, NY	Unknown	53 fat	1	Dioxin and PCBs detected	Watson et al. (1985)

[a] Concentrations reported as TCDD toxic equivalents.
[b] Species listed as *Emydoidea blandingi* in reference.

The highest concentration of total PCDDs reported in field-collected turtles was 596.7 pg/g in the fat of a female snapping turtle collected from the St. Louis River (Ryan et al. 1986). This concentration was very similar to that measured in the fat of a male snapping turtle from the same location. Concentrations of PCDDs in fat were from 4 to 7 times greater than those in the livers of the three turtles. As indicated by Ryan et al. (1986), this ratio is higher than that reported for other chlorinated organics in turtles and is consistent with that found in laboratory mammals (Birnbaum et al. 1985). In the Ryan et al. (1986) study, the distribution of the PCDD congeners in the turtles was similar to that reported in fish collected from the vicinity. A fat sample from a female turtle (species unknown) collected in South Vietnam, however, contained total PCDD and total PCDF concentrations lower than those measured in the ovaries, liver, gall bladder, and muscle (Olie et al. 1989).

In the two studies that measured PCDD concentrations in turtles (Olie et al. 1989; Ryan et al. 1986), 2,3,7,8-tetrachlorodibenzodioxin (2,3,7,8-TCDD) was the predominant form. This PCDD congener has been found to predominate in fish and other wildlife species from contaminated areas (Eisler 1986b).

Data on the biomagnification of PCDDs are inconclusive because of the limited data. The congener 2,3,7,8-TCDD may be biomagnified through the food chain as evidenced by higher concentrations of the compound in snapping turtles than in fish, eels, and herring gulls collected from the St. Lawrence/Lake Ontario area (Ryan et al. 1986).

With reference to the PCDFs, the highest concentration measured in field-collected turtles was 4909 pg/g in a fat sample from a female snapping turtle collected from the St. Lawrence River (Ryan et al. 1986) (Table 4). Concentrations of PCDFs in the fat of the three turtles collected from the river ranged from 5.7 to 7.8 times that measured in liver tissues, similar to that reported for PCDDs in the same turtles. PCDFs in fat tissues of gravid females may be remobilized prior to egg laying and contribute to variability of PCB concentrations in organs and tissues (Olie et al. 1989).

Among the PCDF compounds measured in turtle tissues, 2,3,4,7,8-pentachlorodibenzofuran and 2,3,7,8-tetrachlorodibenzofuran have been found to predominate (Ryan et al. 1986; Rappe et al. 1981; Olie et al. 1989). This is in agreement with that reported in field-collected fish (Stalling et al. 1983) and with the relative contribution of the major PCDFs found in Aroclor 1254 and Aroclor 1260 (Rappe and Buser 1980). Stalling et al. (1983) point out that the isomers with 2,3,7,8-chlorine substitutions are preferentially biomagnified in fish and birds.

Although there are limited data on PCDD and PCDF concentrations in turtles, trends can be noted. PCDDs and PCDFs distribute in turtle tissues in a manner similar to that for PCBs and organochlorine pesticides. Limited data indicate that certain PCDDs, specifically 2,3,7,8-TCDD, may be biomagnified. Finally, the predominant forms of these compounds in turtles are the 2,3,7,8 isomers.

D. Metals

Heavy metals have been monitored extensively in both terrestrial and aquatic systems; however, only a few studies have used turtles as monitors of metal contamination in the environment (i.e., Albers et al. 1986; Stoneburner et al. 1980; Overmann and Krajieck 1989). The metals that have been measured in either field-collected turtles or turtle eggs are lead, mercury, cadmium, chromium, copper, zinc, nickel, molybdenum, iron, cobalt, aluminum, strontium, and barium (Table 5). The majority of the information available focuses on residue concentrations and tissue distributions of these metals without reference to ambient metal concentrations.

Tissue distributions of metals in turtles appear similar to those reported in mammals. Bone and shell contain the highest concentrations of lead, followed by liver and kidney tissues (Beresford et al. 1981; Overman and Krajicek 1989). Carapace and blood samples are recommended for routine monitoring of lead in turtles because these tissues can be obtained without harm to the animal (Overman and Krajicek 1989). Among the soft tissues, the highest concentrations of cadmium were detected in kidney (Robinson and Wells 1975). Mercury, chromium, nickel, and zinc have been measured in the kidneys and livers of turtles (Albers et al. 1986; Helwig and Hora 1983; Flickinger and King 1972; Meyers-Schöne et al. 1993). These tissues usually contain the highest concentrations of metals (Goyer 1991) and are commonly used in the monitoring of metals in wild mammals and birds (Eisler 1985a, 1986a, 1987).

Concentrations of some metals have been found to vary with the gender of the turtle. Among the metals analyzed by Albers et al. (1986), only copper was found to differ between the two sexes. Significantly higher concentrations of the metal were measured in the livers of male snapping turtles than those of females from the same site. Meyers-Schöne et al. (1993) also reported that male sliders generally contained higher concentrations of mercury in both kidney and muscle tissues than females of the same species. Sex-related differences reported by Albers et al. (1986) and Meyers-Schöne et al. (1993) may be attributed to elimination of metals by gravid females into their eggs or differences in exposure due to activity patterns or feeding habits that may vary during the breeding season. Positive correlations between turtle weight and mercury concentration in muscle tissue have also been reported for adult snapping turtles (Meyers-Schöne et al. 1993).

Several metals have been measured in the eggs of loggerhead sea turtles (Stoneburner et al. 1980; Hillestad et al. 1974). Significantly higher concentrations of cadmium, copper, and lead were found in egg yolk than in

Table 5. Metal Concentrations in Field-Collected Turtles

Metal	Species	Location	Sex	Concentration (μg/g wet weight)	N	Observation	Reference
Aluminum	Caretta caretta	National seashores, FL, GA, NC	—	3.56–6.30[a] egg yolks	96	Range of means from 4 sites; highest in eggs from Canaveral National Seashore, FL	Stoneburner et al. (1980)
Barium	Caretta caretta	National seashores, FL, GA, NC	—	2.09–6.87[a] egg yolks	96	Range of means from 4 sites; highest in eggs from Cape Lookout National Seashore, NC	Stoneburner et al. (1980)
Cadmium	Apalone spinifera[b]	River that received effluent from plating industries, TN	F	9.87 kidney 0.19 small intestine	12 12	Highest in kidney and lowest in small intestine	Robinson and Wells (1975)
	Caretta caretta	Seashores, GA, SC	—	0.17[c] egg yolks 0.56[c] albumin	Unknown Unknown	Higher in albumin than yolk	Hillestad et al. (1974)
	Caretta caretta	National seashores, FL, GA, NC	—	0.026–0.20[a] egg yolks	96	Range of means from 4 sites; means >0.1 μg/g (dw) in eggs from Cumberland Island National Seashore, GA, and Canaveral National Seashore, FL	Stoneburner et al. (1980)
	Chelydra serpentina	Contaminated wetland #1 (brackish), NJ	M F M	0.24 kidney 0.10 liver 0.30 kidney 0.08 liver 0.09 kidney 0.08 liver	8 8 3 3 8 8	Low concentrations in all animals; highest in kidney of turtles from NJ brackish water sites (#1)	Albers et al. (1986)
		Contaminated wetland #2 (freshwater), NJ					

	Species	Location	Sex	Value	n	Notes	Reference
	Chelydra serpentina	Reference wetland, MD	M / F	0.07 kidney / 0.07 liver / 0.07 kidney / 0.06 liver	7 / 7 / 6 / 6	Detected in turtles from all sites	Albers et al. (1986)
	Chelydra serpentina	5 river and 1 lake site, MN	M / F	0.010 muscle / 0.012 muscle	8 / 4	Range of values 0.002–0.025 $\mu g/g$	Helwig and Hora (1983)
Chromium	*Caretta caretta*	National seashores, FL, GA, NC	—	1.04–1.71^a egg yolks	96	Range of means from 4 sites; highest in eggs from Cumberland Island National Seashore, GA	Stoneburner et al. (1980)
	Chelydra serpentina	Contaminated wetland (brackish) #1, NJ	M / F	2.97 kidney / 0.60 liver / 2.70 kidney / 0.60 liver	8 / 8 / 3 / 3	Highest in kidney of turtles from NJ brackish water site (#1)	Albers et al. (1986)
		Contaminated wetland (freshwater) #2, NJ	M	1.13 kidney / 0.36 liver	8 / 8		
	Chelydra serpentina	Reference wetland, MD	M / F	0.93 kidney / 1.00 liver / 1.26 kidney / 1.97 liver	7 / 7 / 6 / 6	Detected in turtles from all sites	Albers et al. (1986)
Cobalt	*Caretta caretta*	National seashores, FL, GA, NC	—	0.0–0.73^a egg yolks	96	Range of means from 4 sites; low concentrations	Stoneburner et al. (1980)
Copper	*Caretta caretta*	Seashores, GA, SC	—	2.08^c egg yolks / 6.0^c albumin	Unknown / Unknown	Higher in albumin than yolk	Hillestad et al. (1974)
	Caretta caretta	National seashores, FL, GA, NC	—	4.97–6.61^a egg yolks	96	In eggs from all sites	Stoneburner et al. (1980)

(*Continued*)

Table 5. (Continued)

Metal	Species	Location	Sex	Concentration (μg/g wet weight)	N	Observation	Reference
	Chelydra serpentina	Contaminated wetland (brackish) #1, NJ	M	1.81 kidney	8	Higher in liver than kidney; sex-related difference in NJ brackish water turtles (#1), with concentration in livers of males significantly higher than in females	Albers et al. (1986)
				9.72 liver	8		
			F	1.27 kidney	3		
				5.17 liver	3		
		Contaminated wetland (freshwater) #2, NJ	M	1.73 kidney	8		
				2.08 liver	8		
	Chelydra serpentina	Reference wetland, MD	M	0.82 kidney	7		
				1.28 liver	7		
			F	1.07 kidney	6		
				1.57 liver	6		
Iron	*Caretta caretta*	National seashores, FL, GA, NC	—	71.3–74.7^a egg yolks	96	Range of means from 4 sites; second highest of metals analyzed; similar in eggs from all sites	Stoneburner et al. (1980)
Lead	*Caretta caretta*	Seashores, GA, SC	—	2.87^c egg yolks	Unknown	Higher in albumin than yolk	Hillestad et al. (1974)
				12.0^c albumin	Unknown		
	Caretta caretta	National seashores, FL, GA, NC	—	1.23–2.18^a egg yolks	96	Range of means from 4 sites; highest in eggs from Canaveral National Seashore, FL	Stoneburner et al. (1980)
	Chelydra serpentina	Contaminated wetland (brackish) #1, NJ	M	0.19 kidney	8	Background Hg in turtles from contaminated and reference sites; no consistent relationship between Pb in liver and kidney	Albers et al. (1986)
				nd^d liver	8		
			F	nd kidney	3		
				nd liver	3		
		Contaminated wetland (freshwater) #2, NJ	M	0.10 kidney	8		
				0.12 liver	8		

Chelydra serpentina	Reference wetland, MD	M	0.07 kidney	7	See *Chelydra serpentina*, Albers et al. (1986)
			0.07 liver	7	
		F	0.16 kidney	6	
			nd liver	6	
Chelydra serpentina	Upstream from old lead mining area (area I), MO	M	0.125 muscle	8	In each turtle, bone contained highest Pb followed by carapace. Shell and blood served as useful nonlethal tissue-type samples; progressive increase in tissue lead as sampling moved downstream; male-female differences not noted; concentrations not found to alter capture success or majority of blood indices (d-ALAD levels not reported)
			0.192 brain	8	
			0.182 liver	8	
			0.264 blood	8	
			0.986 carapace	8	
			0.883 bone	3	
		F	0.128 muscle	7	
			0.137 brain	7	
			0.173 liver	7	
			0.298 blood	7	
			0.967 carapace	7	
			1.15 bone	7	
	River inside an old lead mining belt (area II), MO	M	0.322 muscle	9	Overmann and Krajicek (1989)
			0.211 brain	9	
			0.327 liver	9	
			1.00 blood	9	
			13.89 carapace	9	
			48.96 bone	9	
		F	0.158 muscle	5	
			0.150 brain	5	
			0.246 liver	5	
			0.455 blood	5	
			9.54 carapace	5	
			14.76 bone	5	

(Continued)

Table 5. (*Continued*)

Metal	Species	Location	Sex	Concentration (μg/g wet weight)	N	Observation	Reference
		Downstream from an old lead mining area (area III), MO	M	0.209 muscle	4		
				0.311 brain	4		
				0.575 liver	4		
				3.98 blood	4		
				25.19 carapace	4		
				64.83 bone	4		
			F	0.193 muscle	4		
				0.274 brain	4		
				0.405 liver	4		
				0.553 blood	3		
				40.84 carapace	4		
				164.3 bone	4		
	Terrapene carolina	Woodland area near Pb smelter, MO	M	51.8 humerus	4	Highest in bone > kidney > liver	Beresford et al. (1981)
				64.5 femur	4		
				21.6 liver	4		
				24.3 kidney	4		
				6.00 blood	4		
				0.35 skin	4		
				0.20 lung	4		
	Terrapene carolina	Reference woodland, WV	M	4.51 humerus	1	Lead in liver, kidney, skin, blood, and bone significantly higher near smelter; levels similar in liver and kidney	Beresford et al. (1981)
				5.55 femus	1		
				2.21 liver	1		
				4.83 kidney	1		
				0.22 blood	1		
				nd skin	1		
				nd lung	1		

	Species	Location	Sex	Concentration	n	Comments	Reference
Mercury							
	Caretta caretta	Seashores, GA, SC	—	0.02–0.09 egg yolks 0.01–0.03 albumin	Unknown Unknown	Traces in yolk and albumin	Hillestad et al. (1974)
	Caretta caretta	National seashores, FL, GA, NC	—	0.41–1.39[a] egg yolks	96	Range of means from 4 sites; highest in eggs from Cumberland Island National Seashore, GA	Stoneburner et al. (1980)
			F	3.41 humerus	1		
				3.21 femur	3		
				0.80 liver	3		
				0.77 kidney	3		
				nd, 0.15 blood	3		
				nd, 0.16 skin	3		
				nd lung	3		
	Chelydra serpentina	Contaminated wetland #1, (brackish) NJ	M	0.55 kidney	8	Highest in liver; although Hg detected in sediments (0.4–2.80 ppm), relatively low in turtles	Albers et al. (1986)
				1.28 liver	8		
			F	0.41 kidney	3		
				1.27 liver	3		
		Contaminated wetland #2, (freshwater) NJ	M	0.39 kidney	8		
				0.60 liver	8		
	Chelydra serpentina	Reference wetland, MD	M	0.44 kidney	7	Detected in animals from reference site	Albers et al. (1986)
				0.90 liver	7		
			F	0.56 kidney	6		
				0.46 liver	6		
	Chelydra serpentina	5 river and 1 lake site, MN	M	<0.02–0.04 muscle	9	Highest in muscle of females	Helwig and Hora (1983)
				0.12 muscle	10		
				0.03 fat	5		
				0.24 muscle	5		
			F	0.03 fat	1		
			Unknown	0.14–0.30 muscle	2		

(Continued)

Table 5. (*Continued*)

Metal	Species	Location	Sex	Concentration (μg/g wet weight)	N	Observation	Reference
	Chelydra serpentina	Contaminated lake, TN	M	1.30 kidney 0.17 muscle 0.41 kidney	12 12 6	Turtles from contaminated site significantly higher than for reference site; higher in kidney than muscle; significantly higher in *C. serpentina* than *T. scripta* tissues from contaminated site; concentrations correlated with turtle weight	Flickinger and King (1972) Meyers-Schöne (1989) Meyers-Schöne et al. (1993)
		Reference wetland, TN	M F	0.12 muscle 0.21 kidney 0.06 muscle	6 3 3		
	Kinosternon flavescens	Rice field, TX	Unknown	0.12 whole-body^e	3	Low concentrations	Flickinger and King (1972)
	Trachemys scripta	Contaminated lake, TN	M	1.11 kidney 0.15 muscle	6 6	Contaminated site significantly higher than from reference site; higher in kidney than muscle; concentrations correlated with turtle sex; also see *Chelydra serpentina*, Meyers-Schöne (1989)	Meyers-Schöne (1989)
			F	0.18 kidney 0.06 muscle	6 6		
		Reference wetland, TN	M	0.15 kidney 0.03 muscle	6 6		
			F	0.09 kidney 0.03 muscle	6 6		
	Trachemys scripta elegans^f	Rice field, TX	Unknown	0.08 whole-body nd unlaid eggs	2 4	Higher in adults than eggs	Flickinger and King (1972)

Element	Species	Location	Sex	Concentration	n	Comments	Reference
Molybdenum	*Caretta caretta*	National seashores, FL, GA, NC	—	2.67–17.9^a egg yolks	96	Range of means from 4 sites; highest in eggs from Canaveral National Seashore, FL	Stoneburner et al. (1980)
Nickel	*Caretta caretta*	National seashores, FL, GA, NC	—	2.28^a egg yolks	96	Range of means from 4 sites; only eggs from Cape Lookout National Seashore, NC contained >0.26 µg/g (dw)	Stoneburner et al. (1980)
	Chelydra serpentina	Contaminated wetland #1 (brackish), NJ	M	1.24 kidney	8	Highest mean in kidney from NJ brackish water sites (#1)	Albers et al. (1986)
				0.24 liver	8		
			F	1.07 kidney	3		
				0.27 liver	3		
		Contaminated wetland #2 (freshwater), NJ	M	0.45 kidney	8		
				0.13 liver	8		
	Chelydra serpentina	Reference wetland, MD	M	0.35 kidney	7	See above, Albers et al. (1986)	Albers et al. (1986)
				0.44 liver	7		
			F	0.43 kidney	6		
				0.99 liver	6		
Strontium	*Caretta caretta*	National seashores, FL, GA, NC	—	66.1–74.0^a egg yolks	96	Range of means from 4 sites; third highest of metals analyzed	Stoneburner et al. (1980)
Zinc	*Caretta caretta*	Seashores, GA, SC	—	32.25^c egg yolks	Unknown	Slightly higher in egg yolks	Hillestad et al. (1974)
				26^c albumin	Unknown		
	Caretta caretta	National seashores, FL, GA, NC	—	73.5–80.5^a egg yolks	96	Range of means from 4 sites; of the heavy metals, highest concentrations were reported for zinc	Stoneburner et al. (1980)

(Continued)

Table 5. (Continued)

Metal	Species	Location	Sex	Concentration (μg/g wet weight)	N	Observation	Reference
	Chelydra serpentina	Contaminated wetland #1 (brackish), NJ	M	9.93 kidney	8	Highest in liver, and in livers of males from NJ brackish-water site (#1)	Albers et al. (1986)
				50.4 liver	8		
			F	9.79 kidney	3		
				39.0 liver	3		
		Contaminated wetland #2 (freshwater), NJ	M	10.5 kidney	8		
				30.7 liver	8		
	Chelydra serpentina	Reference wetland MD	M	8.80 kidney	7	See above, Albers et al. (1986)	Albers et al. (1986)
				27.7 liver	7		
			F	9.60 kidney	6		
				29.3 liver	6		

[a]Concentration based on dry weight.
[b]Species listed as *Trionyx spiniferus* in reference.
[c]Weight basis (dry or wet) not reported.
[d]nd = Not detectable (detection limit not reported).
[e]All whole-body analyses did not include shell.
[f]Species listed as *Pseudemys scripta elegans* in reference.

albumin (Hillestad et al. 1974). In addition, of the metals analyzed in the eggs, the highest concentrations were of zinc [73.5–80.5 µg/g, dry weight (dw)], iron (71.3–74.7 µg/g, dw), and strontium (66.1–74.0 µg/g, dw) (Stoneburner et al. 1980). The high concentrations of zinc and iron may be attributed to their requirement as an essential metal. Strontium is an analog of calcium, and it may have been transferred to the egg in place of calcium. Because the eggs were collected in nests, concentrations detected in the eggs do not provide conclusive evidence that metals can be transferred to eggs in utero. Because the loggerhead is protected under the U.S. Endangered Species Act, adults cannot be legally killed to determine metal concentrations of eggs in utero. Thus, eggs from nonendangered freshwater turtles collected in utero should be analyzed for metals and compared with oviposited eggs to establish the contribution of the mother to inorganic contaminants in eggs.

The potential for methylmercury to be biomagnified in certain species of aquatic turtles has been suggested. Mercury concentrations were significantly higher in snapping turtles than in sliders collected from the same location (Meyers-Schöne 1989; Meyers-Schöne et al. 1993). This difference was found to be consistent with differences in the food habits of the two species.

Several general trends emerge from these studies of metals in turtles. A large variety of metals have been measured in both turtle tissues and eggs. The highest tissue concentrations of most metals occur in kidney and liver. Lead concentrations are, however, greater in bone and carapace. Concentrations of metals in turtles may vary with gender, as evidenced by mercury and copper. Species differences have been reported for mercury and related to differences in food habits; biomagnification of mercury is implicated by the data.

E. Radionuclides

Radionuclide contaminants, whether released to the environment as fallout products from nuclear weapons testing, as nuclear waste, or as deliberate releases from isotope tracer studies, have been detected in both freshwater and terrestrial turtles. An excellent comprehensive review (Hinton and Scott 1990) of radionuclides in reptiles and amphibians is available; the present review emphasizes radionuclides as they pertain to the use of turtles as biological monitors of environmental contamination. The radionuclides that have been reported in field-collected turtles are ^{90}Sr, ^{137}Cs, ^{60}Co, ^{85}Sr, ^{85}Zn, and ^{131}I (Table 6).

Strontium-90 concentrations have been measured in whole turtles, bones, and shells (Scott et al. 1986; Towns 1987; Holcomb et al. 1971; Jackson et al. 1974; Meyers-Schöne 1989; Meyers-Schöne et al. 1993). Dissection and analysis of various tissues revealed that 99% of the whole-body burden of ^{90}Sr was contained in the shell and bone (Towns 1987). Elimina-

Table 6. Radionuclide Concentrations in Field-Collected Turtles

Radionuclide	Species	Location	Sex	Concentration (Bq/g wet weight)[a]	N	Observation	Reference
^{137}Cs	Chelydra serpentina	Spiked pond, OH	Unknown	1.7–3.5^b whole-body[c]	2	1.48×10^8 Bq ^{137}Cs added to pond; turtles were immigrants	Brungs (1967)
	Chelydra serpentina	Contaminated lake, TN	M	17.4×10^{-2} liver	12	Turtles from contaminated site above background; not detected in turtles from reference site	Meyers-Schöne (1989)
				39.6×10^{-2} muscle	12		Meyers-Schöne et al. (1993)
		Reference wetland, TN	M	$<3.7 \times 10^{-3}$ liver	6		
				$<3.7 \times 10^{-3}$ muscle	6		
			F	$<3.7 \times 10^{-3}$ liver	3		
				$<3.7 \times 10^{-3}$ muscle	3		
	Trachemys scripta	Contaminated lake, TN	M	11.6 liver	6	Highest in muscles of two immigrant male T. scripta; muscles higher than liver	Meyers-Schöne (1989)
				89.4 muscle	6		Meyers-Schöne et al. (1993)
			F	7.49×10^{-2} liver	6		
				27.3×10^{-2} muscle	6		
		Reference wetland, TN	M	$<3.7 \times 10^{-3}$ liver	6		
				$<3 \times 10^{-3}$ muscle	6		
			F	$<3 \times 10^{-3}$ liver	6		
				$<3 \times 10^{-3}$–1.44×10^{-2} muscle	6		
	Trachemys scripta[d]	Savannah River Plant seepage basins, SC	Both	9.0 whole-body	15	Initial total body burden reported; 213–643 d determined as time to reach equilibrium; elimination rate constant averaged 0.0072 d^{-1}	Peters and Brisbin (1988) Peters (1986)
	Trachemys scripta[d]	Savannah River Plant seepage basins, SC	Both	2.77 whole-body	36	Initial total body burden reported, seasonal variation in elimination rates; average yearly biological half-life 64 d	Scott et al. (1986)

Isotope	Species	Location	Sex	Concentration/tissue	n	Comments	Reference
	Trachemys scripta[d]	Reference ponds, SC	Both	0.0074 whole-body	26	Contained low levels	Scott et al. (1986)
	Trachemys scripta[d]	Savannah River Plant seepage basin, SC	Unknown	5.79 whole-body	7	Concentrations in organs and tissues were normalized to that in muscle; muscle was twice that in other soft tissues; actual concentrations not reported	Towns (1987)
	Trachemys scripta[d]	Contaminated reservoir, SC	Unknown	0.71 whole-body	10	See above, Towns (1987)	Towns (1987)
⁶⁰Co	Chelydra serpentina	Spiked pond, OH	Unknown	0.77–1.1[b] whole-body	2	1.48×10^8 Bq ⁶⁰Co added to pond; turtles were immigrants	Brungs (1967)
	Chelydra serpentina	Contaminated lake, TN	M	4.76×10^{-2} liver	12	Turtles from contaminated site above background and significantly higher in liver than tissues from reference site	Meyers-Schöne (1989)
			M	$<3.7 \times 10^{-3}$ muscle	12		Meyers-Schöne et al. (1993)
		Reference wetland, TN	M	$<3.7 \times 10^{-3}$ liver	6		
			M	$<3.7 \times 10^{-3}$ muscle	6		
			F	$<3.7 \times 10^{-3}$ liver	3		
			F	$<3.7 \times 10^{-3}$ muscle	3		
	Chelydra serpentina	Lake contaminated with ⁶⁰Co from a nearby disposal pit, Ontario, Canada	Unknown	0.034 whole-body	1	0.37 Bq/L in lake water; concentration factors were calculated	Ophel and Fraser (1971)
				0.14 carapace	1		
				0.004 whole-body minus carapace	1		
	Trachemys scripta	Contaminated lake, TN	M	7.65×10^{-2} liver	6	Highest in livers of two immigrant male T. scripta; interspecies differences not found	Meyers-Schöne (1989)
			M	$<3.7 \times 10^{-3}$ muscle	6		Meyers-Schöne et al. (1993)
			F	4.42×10^{-2} liver	6		
			F	$<3.7 \times 10^{-3}$ muscle	6		
		Reference wetland, TN	M	$<3.7 \times 10^{-3}$ liver	6		
			M	$<3.7 \times 10^{-3}$ muscle	6		
			F	$<3.7 \times 10^{-3}$ liver	6		
			F	$<3.7 \times 10^{-3}$ muscle	6		

(Continued)

Table 6. (Continued)

Radionuclide	Species	Location	Sex	Concentration (µg/g wet weight)	N	Observation	Reference
^{90}Sr	Chelydra serpentina	Uncontaminated sites, MS	Unknown	0.29–1.29^e shell	2	^{90}Sr detected	Holcomb et al. (1971)
	Chelydra serpentina	Uncontaminated sites, FL	Unknown	0.15–0.60^e shell	2	Higher in C. serpentina from GA sites	Jackson et al. (1974)
	Chelydra serpentina	Uncontaminated sites, GA	Unknown	1.9–4.4^e shell	2		
	Chelydra serpentina	Contaminated lake, TN	M	16.5 bone	12	Significantly higher in bone and carapace from contaminated site than those from reference site; bone and carapace ^{90}Sr in a given animal similar; statistically significant differences not found between C. serpentina and T. scripta	Meyers-Schöne (1989)
			M	16.6 carapace	12		Meyers-Schöne et al. (1993)
		Reference wetland, TN	M	$<3.7 \times 10^{-2}$–15 $\times 10^{-2}$ bone	6		
			F	$<3.7 \times 10^{-2}$–15 $\times 10^{-2}$ carapace	3		
				$<3.7 \times 10^{-2}$–17.5 $\times 10^{-2}$ bone	3		
				$<3.7 \times 10^{-2}$–8.25 $\times 10^{-2}$ carapace	3		
	Deirochelys reticularia	Uncontaminated site, FL	Unknown	0.64^e shell	3	^{90}Sr detected	Jackson et al. (1974)
	Gopherus polyphemus	Uncontaminated site, AL	Unknown	4.77^e shell	1	Turtle had second highest concentration of all Species with the highest ^{90}Sr; explanation based on herbivorous diet	Holcomb et al. (1971)
	Gopherus polyphemus	Uncontaminated sites, FL	Unknown	3.6–5.4^e shell	2		Jackson et al. (1974)
		Uncontaminated sites, GA	Unknown	6.46^e shell	6		
		Uncontaminated sites, AL	Unknown	3.57^e shell	3		
	Kinosternon baurii^f	Uncontaminated sites, FL	Unknown	0.67^e shell	3	^{90}Sr detected	Jackson et al. (1974)

Species	Site		Value	N	Finding	Reference
Kinsoternon subrubrum	Uncontaminated sites, MS	Unknown	1.31^e shell	8	No correlation between size and concentration	Holcomb et al. (1971)
Kinosternon subrubrum	Uncontaminated sites, FL	Unknown	1.44^e shell	7	Higher concentrations in *K. subrubrum* than *K. bauri*	Jackson et al. (1974)
Malaclemys terrapin macrospilota	Uncontaminated sites, GA	Unknown	1.63^e shell	1	Low concentration	Jackson et al. (1974)
Pseudemys concinna^g	Uncontaminated site, FL	Unknown	0.02^e shell	1	Higher than in subspecies from FL	Holcomb et al. (1971)
Pseudemys concinna suwanniensis^h	Uncontaminated sites, MS	Unknown	1.76^e shell	3	Low concentration	Jackson et al. (1974)
Pseudemys floridana peninsularis^i	Uncontaminated site, FL	Unknown	0.04^e shell	3	No correlation between size and concentration	Holcomb et al. (1971)
Pseudemys floridana peninsularis^i	Uncontaminated sites, FL	Unknown	0.03^e shell	9	Minus one individual from another site mean = 0.36 Bq/g	Jackson et al. (1974)
Pseudemys nelsoni^j	Uncontaminated sites, FL	Unknown	0.98^e shell	2	Higher than other *Pseudemys*	Jackson et al. (1974)
Sternotherus minor minor	Uncontaminated site, FL	Unknown	1.78^e shell	2	No correlation between size and concentration	Holcomb et al. (1971)
Sternotherus minor minor	Uncontaminated sites, FL	Unknown	0.02^e shell	5	Low levels	Jackson et al. (1974)
Sternotherus odoratus	Uncontaminated sites, FL	Unknown	0.04^e shell	2	Higher in MS turtle	Holcomb et al. (1971)
Sternotherus odoratus	Uncontaminated sites, MS	Unknown	0.05^e shell	1		
Sternotherus odoratus	Uncontaminated sites, FL	Unknown	1.0^e shell	8	Higher in GA population	Jackson et al. (1974)
	Uncontaminated sites, GA	Unknown	0.06^e shell	14		
			1.21^e shell			

(*Continued*)

Table 6. (Continued)

Radionuclide	Species	Location	Sex	Concentration (µg/g wet weight)	N	Observation	Reference
	Terrapene carolina	Uncontaminated sites, MS	Unknown	1.0^e shell	31	Inverse correlation between size and concentration in MS and TN populations	Holcomb et al. (1971)
		Uncontaminated sites, TN	Unknown	1.14^e shell	11		
		Uncontaminated sites, GA	Unknown	1.48^e shell	7		
		Uncontaminated sites, AR	Unknown	0.73^e shell	3		
	Terrapene carolina	Uncontaminated sites, FL	Unknown	0.42–1.1^e shell	2	^{90}Sr detected	Jackson et al. (1974)
	Trachemys scripta^d	Uncontaminated sites, MS	Unknown	0.78^e shell	29	No correlation between size and concentration; Northeastern MS highest concentration among turtles sampled	Holcomb et al. (1971)
			Unknown	10.4^e shell	1		
	Trachemys scripta	Contaminated lake, TN	M	837.5 bone	6	Significantly higher in bone and carapace from contaminated site than those from reference site; highest in two immigrant males; in general, bone and carapace ^{90}Sr in a given animal were similar; also see Chelydra serpentina, Meyers-Schöne (1989)	Meyers-Schöne (1989)
			F	713 carapace	6		Meyers-Schöne et al. (1993)
				15.0 bone	6		
				19.0 carapace	6		
		Reference wetland, TN	M	<3.7 × 10^{-2}–12 × 10^{-2} bone	6		
			F	<3.7 × 10^{-2}–16 × 10^{-2} carapace	6		
				<3.7 × 10^{-2}–23 × 10^{-2} bone	6		
				<3.7 × 10^{-2}–4.7 × 10^{-2} carapace	6		

	Species	Location		Concentration	n	Comments	Reference
	Trachemys scripta[d]	Savannah River Plant seepage basins, SC	Both	94.3 whole-body	36	Initial total body burden reported; seasonal variation in elimination rates; average yearly biological half-life 365 d	Scott et al. (1986)
	Trachemys scripta[d]	Reference ponds, SC	Both	0.39 whole-body	26	Low levels	Scott et al. (1986)
	Trachemys scripta[d]	Savannah River Plant seepage basins, SC	Unknown	130.7 whole-body	7	99% of whole-body concentration attributed to shell and bone	Towns (1987)
	Trachemys scripta[d]	Contaminated reservoir, SC	Unknown	1.37 whole-body	10	See above, Towns (1987)	Towns (1987)
^{85}Sr	*Chelydra serpentina*	Spiked pond, OH	Unknown	4.4–14.4[a] whole-body	2	1.48×10^8 Bq ^{85}Sr added to pond; turtles were immigrants	Brungs (1967)
^{65}Zn	*Chelydra serpentina*	Spiked pond, OH	Unknown	2.8–2.9[a] whole-body	2	1.48×10^8 Bq ^{65}Zn added to pond; turtles were immigrants	Brungs (1967)

[a] pCi/g converted to Bq/g by dividing by 27.
[b] Radionuclide concentration based on dry weight.
[c] Whole-body, in all cases, includes shell.
[d] Species listed as *Pseudemys scripta* in reference.
[e] Radionuclide concentration based on ashed shell weight.
[f] Species listed as *Kinosternon bauri palmarum*.
[g] Species listed as *Chrysemys floridana hoyi* in reference.
[h] Species listes as *Chrysemys concinna suwanniensis* in reference.
[i] Species listed as *Chrysemys floridana penisularis* in reference.
[j] Species listed as *Chrysemys nelsoni* in reference.

tion rates for ^{90}Sr in pond sliders were seasonal and were highest during the spring breeding season. The average biological half-life of ^{90}Sr in pond sliders is approximately 365 days (Scott et al. 1986).

Turtle shells collected from numerous locations throughout the southeastern U.S. show an inverse correlation detected between size (age) and ^{90}Sr concentration in box turtles from Mississippi and Tennessee (Holcomb et al. 1971). An inverse correlation of age with ^{90}Sr concentration was also reported for humans (Kulp 1965) and deer (Farris et al. 1969). Rapid growth and deposition of calcium and its chemical analog strontium into bone and shell of juvenile turtles may result in high concentrations of ^{90}Sr in juveniles. No correlation between age and ^{90}Sr concentration was found in box turtles collected from two other states (Holcomb et al. 1971).

Differences in ^{90}Sr concentrations between turtle species collected at the same sites have been reported and may be related to differences in food habits. A comparison of ^{90}Sr in the herbivorous gopher tortoise to that in the omnivorous box turtle revealed higher concentrations of the radionuclide in the former species (Holcomb et al. 1971). A trophic-level difference in ^{90}Sr concentration was, however, not observed between the snapping turtle, a carnivorous species, and the slider, a herbivorous species (Meyers-Schöne et al. 1993).

Strontium-90 in shells of turtles sampled in several areas throughout the southeastern U.S. was equivalent to or exceeded concentrations of ^{90}Sr in the bones of black-tailed jackrabbits (*Lepus californicus*) collected 32 km north of the Yucca Flat Nevada Test Site at Groom Valley (Holcomb et al. 1971; Jackson et al. 1974). The only known environmental source of ^{90}Sr was from nuclear fallout. The ability of turtles to accumulate and apparently concentrate ^{90}Sr from these areas suggests that the shells of turtles may be a very sensitive and useful source of information on the presence of this radionuclide in the environment.

Cesium-137 has also been studied in turtles (Brungs 1967; Scott et al. 1986; Peters 1986; Towns 1987; Peters and Brisbin 1988; Meyers-Schöne et al. 1993). Analysis of various tissues from sliders collected from an area contaminated with ^{137}Cs revealed that ^{137}Cs was distributed throughout the body; however, the highest concentration of the radionuclide was in muscle (Towns 1987; Meyers-Schöne et al. 1993). The distribution of ^{137}Cs in turtles is similar to that reported in mammals (Stribling et al. 1986). Seasonal variations have been reported for ^{137}Cs in sliders (Scott et al. 1986). Cesium-137 elimination rates were highest during the spring through early summer months, which corresponded to the breeding season and period of highest metabolic activity of the species (Scott et al. 1986). The elimination rate of ^{137}Cs during the spring and summer months was approximately 7.2 ± 4.4 Bq/d (Peters and Brisbin 1988). Positive correlations exist between the weight of the turtle and the elimination rate, and between the weight and the ^{137}Cs concentration in the turtle (initial body burden) (Peters and Brisbin 1988). No relationship was found between the ^{137}Cs elimination rate and turtle gender (Peters and Brisbin 1988), which is consistent with data on birds (Fendley et al. 1977). Because ^{137}Cs is a chemical analog of potassium and not concentrated in hard tissues, ^{137}Cs would be expected to have a much faster elimination rate than ^{90}Sr and a shorter biological half-life. The biological half-life of ^{137}Cs in sliders is 64 d (Scott et al. 1986), several times less than the biological half-life of ^{90}Sr. The time required for pond sliders to reach an equilibrium radiocesium concentration with the environment has been calculated to be 320 d (Peters and Brisbin 1988). A comparison of ^{137}Cs in turtles to that of several other taxa revealed that the biological half-life was greater in turtles than birds (Fendley et al. 1977; Halford et al. 1983) and wild mammals (Jenkins et al. 1969). The longer retention of ^{137}Cs in turtles is consistent with the slow metabolic rate in the poikilothermic turtles. In general, the biological half-life of ^{137}Cs is longer in poikilotherms than in homeotherms (Mailhot et al. 1989).

Only limited information exists on other radionuclides in turtles. An inverse relationship was reported to exist between the concentration of ^{60}Co in an organism and its trophic level. Ophel and Fraser (1971) compared the concentration of ^{60}Co in water (filtered with a 1.2-μm filter) with that in plants, fish, reptiles, and amphibians from a contaminated pond. Concentration factors in aquatic plants relative to water ranged from 270 to 2790 times, whereas concentrations in whole fish exceeded those in water by 9–130 times. Cobalt-60 in a whole snapping turtle was 90 times as great as the concentration in water. No difference in ^{60}Co concentrations was detected between snapping turtles and sliders collected from a lake containing radionuclide contaminants (Meyers-Schöne et al. 1993).

Turtles have been shown to concentrate ^{131}I into thyroid tissue (Shellabarger et al. 1956; Gibbs et al. 1964). In a laboratory study, adults were shown to retain a greater amount of the original dose than did juveniles (Gibbs et al. 1964). Among the vertebrates that inhabit wetlands, turtles (specifically adult turtles) may prove most useful in the assessment of radioiodine contamination. Turtles are generally long-lived and have well-developed thyroid glands, whereas fish do not. These factors suggest that turtles may be a more effective indicator of ^{131}I in the environment than fish. Because turtles are likely to have greater exposure to contaminants in water and sediment than birds or mammals in wetlands, turtles may be highly effective as monitors of radioiodine contamination in wetlands.

Although only a few radionuclides have been analyzed in turtles, information exists that is of value to the use of turtles as biomonitors. Strontium-90 has been found to concentrate primarily in bone and shell. Turtle shells provide a good nondestructive tissue for the environmental monitoring of ^{90}Sr. Cesium-137 concentrations are highest in muscle tissue, ^{60}Co concentrates in kidney and liver tissues, and ^{131}I concentrates in thyroid tissue. Gender differences have not been reported for these radionuclides. Finally, data on interspecific differences in specific radionuclide concentrations are inconclusive.

IV. Other Biomonitoring Tools

The measurement of specific biochemical and histopathological parameters can serve as supplemental or alternative tools in the assessment of exposure to contaminants in field-collected turtles. Although residue concentrations can indicate whether a species has been exposed to contaminants in the environment, they do not yield information on the response of the species to a contaminant. Information on the use of biochemical, histopathological, and physiological indicators of turtle exposure to chemical contamination in the field is summarized below.

A. Biochemical and Histopathological Responses to Stress

Few studies have been published on biochemical and tissue responses of turtles to chemical and physical perturbations in the field. Laboratory studies have shown turtles to be more tolerant to radiation than many other vertebrate species (Cosgrove 1965; Cosgrove 1971). Atland et al. (1951) investigated the effects of whole-body irradiation on eastern box turtles with x-ray exposures ranging from 0.129 to 2.58 C/kg (500–10,000 R). Turtles irradiated with as much as 0.258 C/kg (1000 R) showed 100% survival for 4 mon, suggesting their ability to withstand higher acute doses than birds (Abraham 1972) and mammals (Dunaway et al. 1969). A partial explanation for the higher radiation tolerance of turtles may be related to the shielding effect of the shell, which can reduce the average internal exposure by 21% (Cosgrove 1965). The low metabolic rate of turtles compared with that of mammals may also be a factor. LD_{50} values have been determined for turtles exposed to various intensities of x-rays. The LD_{50} values for three species of turtles after 120-d exposure were reported to range from 0.219 (Cosgrove 1965) to 0.267 C/kg (850–1035 R) for box turtles, to slightly less than 0.258 C/kg (1000 R) for juvenile painted turtles and sliders, to <0.206 C/kg (800 R) for juvenile snapping turtles (Cosgrove 1971). The principal effects observed were on blood-forming tissues and reproductive organs (Atland et al. 1951; Cosgrove 1965). Species comparisons of responses to radiation indicate that juvenile snapping turtles may be more sensitive than juveniles of two other turtle species; thus, snapping turtles may be the turtle species of choice to detect genotoxic damage caused by radiation.

Field studies have revealed that blood and blood plasma characteristics may not be useful indicators of exposure to chemical contaminants in turtles (Albers et al. 1986; Overmann and Krajicek 1989). According to Albers et al. (1986), differences in the levels of protein, albumin, and plasma glucose in individual snapping turtles may be attributed to differences in age. Moreover, plasma glucose levels may be elevated by the stress to individual turtles as a result of collection and handling. In addition, variations in the concentration of hemoglobin can be attributed to differences in the salinities of the waters from which the turtles were captured. Overmann and Krajicek (1989) also found that hematocrit, plasma osmolality, and plasma chlorine ion content were not sensitive indicators of lead exposure in snapping turtles containing blood lead levels of up to 0.85 µg/g. Blood glucose data proved inconclusive.

In addition to research conducted on organochlorine pesticides and turtles, parathion poisoning has been studied in Caspian terrapins (*Mauremys caspica rivulata*) in the laboratory (Yawetz et al. 1983). An LD_{50} of 15 ppm was found for Caspian terrapins exposed to parathion. This concentration is between the LD_{50} range of 5 and 50 ppm reported for most higher vertebrates. It was further shown that the ratio of the rate of activation of parathion to paraoxon to the rate of detoxication of paraoxon to p-nitrophenol was >1 (Yawetz et al. 1983) and approximately twice that reported in barn owls (Yawetz et al. 1979). According to Yawetz et al. (1983), turtles show low inhibition of brain acetylcholinesterase, which is from two to three orders of magnitude lower than that measured in birds. It has been suggested that the reduced enzyme affinity may serve to protect turtles against parathion poisoning. This suggests that acetylcholinesterase inhibition in turtles may not be a good indicator of exposure to organophosphate pesticides.

The induction of mixed-function oxidase (MFO) systems has been examined in Caspian terrapins exposed to PCBs (Yewetz et al. 1983). The MFO system was not induced in liver tissues of individual turtles following oral exposure to six treatments of 125 ppm each of Aroclor 1254 over a period of 3 weeks. Caspian terrapins collected from a canal contained PCB liver concentrations of approximately 23 mg/kg.

Measurements of the DNA content in sliders collected from seepage basins containing radioactive and nonradioactive contaminants and from a reference site proved to be a useful indicator of exposure to genotoxic agents in the environment (Bickham et al. 1988). Bickham et al. (1988) reported that the only turtles to contain mosaic DNA were four individuals from the contaminated site. A comparison of the coefficients of variation in DNA of turtles with normal DNA histograms (nonmosaics) revealed a significant difference between the DNA of turtles from the contaminated and reference sites. The higher coefficient of variation in turtles from the seepage basins may have resulted from mutations (deletions or duplications) in DNA induced by radiation and/or chemical agents (Bickham et al. 1988). In a related study, Lamb et al. (1991) reported elevated coefficients of variation in the DNA of sliders containing mean ^{137}Cs and ^{90}Sr body burdens of 842 and 1879 Bq, respectively. These turtles were collected from an inactive nuclear reactor cooling reservoir containing radionuclides. Their findings provide evidence that ionizing radiation can induce changes in turtle DNA as measured by flow cytometry. In addition to the DNA changes noted by Bickham et al. (1988), a positive correlation was shown between the coefficient of variation and plastron length in male turtles. The investigators proposed that larger, older turtles may be better indicators of environmentally induced mutagenic effects than younger turtles or shorter-lived species because periods of exposure are larger and the effects of mutations may accumulate over time.

Genotoxic damage was also measured in turtles collected from a settling basin for radioactive and nonradioactive contaminants using a DNA alkaline unwinding assay (Meyers-Schöne et al. 1993). Both species studied, sliders and snapping turtles, were found to contain a significantly higher frequency of DNA breaks than did the same species of turtles from the reference site. The specific causative agent was not identified; however, the estimated ^{137}Cs and ^{90}Sr body burdens of the turtles were within the range of ^{137}Cs and ^{90}Sr body burdens reported by Lamb et al. (1991) in turtles with genetic damage (Meyers-Schöne et al. 1993).

B. Turtle Growth Rates in Relation to Contamination

Growth rates have been measured in turtles; however, differences in the growth of individual turtles cannot be reliably attributed to contaminants in the environment. Growth rates may vary with age as noted by Gibbons (1968) and Kiviat (1980), who observed a lower growth rate in adult snapping turtles compared with that of juveniles collected from uncontaminated sites. Differences in growth rates may also be a response to the ingestion of nutrient-deficient food items by turtles (Albers et al. 1986). In addition, the diets of omnivorous species can fluctuate with the availability of foods, which results in higher growth rates in turtles with protein-rich diets (Knight and Gibbons 1968; Graham and Perkins 1976). In one study, the growth rates of four box turtles collected from a DDT-contaminated site were measured and compared with the growth rates of five box turtles from a reference site and found to be very similar (Stickel 1951). The aerial application of DDT (from 1.24 to 2.25 kg/ha) over a 4-yr period did not impair the growth of individual turtles in this study. Because of the lack of data to show that growth is a sensitive indicator of toxicant exposure and the numerous other factors known to affect turtle growth, growth cannot be recommended as evidence of chemical intoxication.

V. Turtles as Monitors: Advantages and Limitations

Turtles possess several attributes that make them potentially useful as monitors of contaminated freshwater and terrestrial environments. Not only are turtles commonly found throughout North America, but they also occupy a variety of habitat types and are relatively easy to capture as well. They are long-lived and may reach an age of 10–50 yr in the wild (Gibbons 1987). This allows for long-term exposure to contaminants. Many species travel distances of a few hundred meters to several kilometers during a year. The fact that turtles are both long-lived and mobile allows for the integration of exposure over time and space. In addition to these attributes, key tissues useful in residue and biochemical analyses are available in large quantities from most turtle species. On the basis of tissue residue concentrations, turtles have been shown to be excellent monitors of PCBs and ^{90}Sr. The more carnivorous species have been found to be effective monitors of contaminants, such as mercury, that biomagnify through food chains. Finally, positive responses have been reported on the use of DNA damage in turtles exposed to environments containing mixtures of radioactive and nonradioactive contaminants.

In addition to the destructive sampling and analysis of turtles, alternatives exist to using the soft tissues in order to monitor the accumulation of certain environmental contaminants in turtles. Because lead and ^{90}Sr concentrate in bone and shell, samples of marginal scutes of the carapace can be removed to assay for chemicals without harm to the turtle. A few milliliters of blood may also be removed from a turtle and analyzed for activities of enzymes that respond to contaminants (e.g., δ-aminolevulenic acid dehydrase and acetylcholinesterase) or to obtain DNA for analysis of genotoxic damage. In addition, most contaminants detected in adults may also be present in the eggs.

Limitations to the use of turtles focus around the capture, handling, and lack of sufficient data on the use of specific biochemical indicators of contaminant exposure. Depending on the species, it may be difficult to collect sufficient numbers of a particular species to constitute an acceptable sample size. Thus, the selection of species is of paramount importance in biological monitoring programs. Biochemical indicators of stressed environments have been developed for mammals, birds, and fish. Existing studies on the use of biochemical indicators of toxicant exposure in turtles are promising. Additional studies, however, are needed to fully evaluate the utility of turtle biomarkers.

Summary

This review was conducted to evaluate turtles as monitors of chemical contamination in the environment. Species-specific differences, habitat use, and trophic relations are identified as important to the selection of an appropriate turtle species for use as an indicator of exposure to specific chemical contaminants. Tissue distributions of organic chemicals, inorganic chemicals, and radionuclides in turtles are similar to those reported in mammals and birds. Organochlorine pesticides, PCBs, PCDDs, and PCDFs accumulate primarily in adipose tissue; metals concentrate primarily in liver and kidney tissues, except for lead, which concentrates in bone and shell. As for the radionuclides, the highest tissue concentrations of ^{137}Cs are found in muscle, whereas ^{90}Sr accumulates almost exclusively in calcified tissue. Samples of turtle shell can serve as a nondestructive method for the analysis of lead and strontium concentrations in field-collected turtles.

The transfer of contaminants from the female turtle to her eggs occurs for PCBs, DDT, and metabolites of DDT. Thus, turtle eggs can serve as indicators of exposure to toxicants under field conditions. Female turtles may have lower concentrations of chlorinated organic compounds in their muscle and fat tissues during the breeding season than male turtles. Therefore, it may be wise to use males during these periods.

Biomagnification of xenobiotic chemicals in turtles has been shown only for PCBs and mercury. Concentrations of PCBs in turtles often exceed those reported to occur in mammals and birds in the same sampling area. Moreover, turtles appear to withstand high concentrations of PCBs without apparent adverse effects on the individual turtles. Consequently, turtles appear to be excellent monitors of PCBs. Mercury concentrations have been correlated with the food habits of turtles, and more carnivorous species contain higher concentrations of mercury.

Few studies have been performed to examine the usefulness of biochemical or physiological parameters as indicators of turtle exposure to chemical contaminants. Findings of alterations in the DNA of turtles inhabiting radioactively contaminated sites encourage the use and testing of other biomarkers on turtles in the wild. Additional research is required to evaluate whether measurements of acetylcholinesterase and MFO enzymes are sensitive indicators of exposure to specific contaminants. Growth has also been measured in turtles as an indicator of exposure to chemical contaminants, but is neither a sensitive nor a chemically specific indicator of toxicant exposure.

Tolerance to high concentrations of radionuclides, PCBs, and parathion has been reported in field-collected turtles. Turtles appear to be more resistant to external radiation than mammals and amphibians, which may be attributed to the shielding effects of the shell. A higher tolerance of turtles to concentrations of PCBs and parathion than in mammals and birds may be related to differences in biochemical detoxification responses. These characteristics favor the use of turtles as monitors of these agents.

Four species of turtles are frequently used to detect contaminants in the environment: the box turtles (*Terrapene carolina* and *T. ornata*), snapping turtles, and sliders. Box turtles are omnivorous and thus have exposure to contaminants that concentrate in both plant (e.g., ^{90}Sr) and animal matter (e.g., chlorinated organics). They are common to most parts of the central and eastern U.S., and their longevity ensures chronic exposure to many of the contaminants in terrestrial environments. For these reasons, the box turtle is recommended as the turtle of choice for terrestrial biological monitoring studies.

The snapping turtle and slider are most often used to monitor contaminants in freshwater environments. The use of snapping turtles as biomonitors has several advantages in addition to the availability of a comparative database for the species. They are large turtles; thus, adequate tissue samples are readily obtained for residue analyses. These animals are aggressive yet easy to trap. Snapping turtles are omnivorous, ingesting large portions of animal matter, which makes them especially useful in the monitoring of compounds such as PCBs, organochlorine pesticides, and methylmercury, which can be biomagnified through food chains. Snapping turtles, however, may not be abundant in small bodies of water. If small ponds or ditches are of interest, mud turtles may be used, but these species are generally difficult to trap.

Sliders have been used frequently in radionuclide investigations. They are easily trapped and very common in ponds and shallow bodies of freshwater. Because adults are primarily herbivorous, they are expected to be especially useful monitors of those chemicals that are not biomagnified.

The USEPA has included two species of turtles in its *Wildlife Exposure Factors Handbook* (in press): the snapping turtle and painted turtle. Although more omnivorous than the slider, the painted turtle has a wider geographic distribution in the U.S. The painted turtle is easily trapped and is likely to serve as an equally useful monitor of contaminants in freshwater environments. Its omnivorous nature may also render it useful for the monitoring of those compounds that can be biomagnified. In general, the use of snapping turtles, sliders, or painted turtles can be recommended to monitor contaminants in freshwater ecosystems. The species of choice for a particular situation, however, depends on the chemical contaminant of concern and local abundance of the species.

Because sea turtles are protected by law, they should not be used to monitor chemical and radionuclide contaminants in marine environments. Analyses of eggs from green sea turtles and loggerheads show the presence of chlorinated organic compounds, specifically DDT and PCBs, in the open seas. Because sea turtles are so long-lived and highly mobile, residue data for sea turtles are of limited utility in identifying point sources of chemical contamination to oceans.

Regardless of whether the species of turtle is aquatic or terrestrial, the state Game and Fish Department and regional U.S. Fish and Wildlife Service office should be consulted to confirm that the species selected for monitoring is not endangered, threatened, or protected by any state or federal regulations. Efforts should also be made to obtain appropriate collection permits. Finally, in order to prevent the oversampling of a species from a given area, the relative abundance of the indicator species should be known prior to any destructive sampling.

Acknowledgments

We thank J.M. Loar, S.S. Talmage, and T.L. Ashwood, Oak Ridge National Laboratory; A.C. Echternacht, University of Tennessee; and J.T. Collins, University of Kansas, for their contributions to this study. The work was conducted in the Environmental Sciences Division at Oak Ridge National Laboratory under the sponsorship of the Office of Environmental Restoration and Waste Management, U. S. Department of Energy, under

Contract DE-AC05-84OR21400 with Martin Marietta Energy Systems, Inc. Publication No. 4154, Environmental Sciences Division, ORNL.

References

Abraham RL (1972) Mortality of mallards exposed to gamma radiation. Radiat Res 49:322-327.

Albers PH, Sileo L, Mulhern BM (1986) Effects of environmental contaminants on snapping turtles of a tidal wetland. Arch Environ Contam Toxicol 15:39-49.

Atland PD, Highman B, Wood B (1951) Some effects of x-irradiation on turtles. J Exp Zool 118:1-19.

Beresford WA, Donovan MP, Henninger JM, Waalkes MP (1981) Lead in the bone and soft tissues of box turtles caught near smelters. Bull Environ Contam Toxicol 27:349-352.

Bickham JW, Hanks BG, Smolen MJ, Lamb T, Gibbons JW (1988) Flow cytometry analysis of the effects of low-level radiation exposure on natural populations of slider turtles (*Pseudemys scripta*). Arch Environ Contam Toxicol 17:837-841.

Birnbaum LS, Weber H, Harris MW, Lamb IV JC, McKinney JD (1985) Toxic interactions of specific polychlorinated biphenyls and 2,3,7,8-tetrachlorodibenzo-p-dioxin:Increased incidence of cleft palate in mice. Toxicol Appl Pharmacol 77:292-302.

Brown MP, Werner MB, Sloan RJ, Simpson KW (1985) Polychlorinated biphenyls in the Hudson River. Environ Sci Technol 19:656-661.

Brungs WA (1967) Distribution of cobalt 60, zinc 65, strontium 85, and cesium 137 in a freshwater pond. US Dept of Health, Education, and Welfare, Public Hlth Service, Nat Ctr for Radiological Health, Rockville, MD.

Bryan AM, Olafsson PG, Stone WB (1987a) Disposition of low and high environmental concentrations of PCBs in snapping turtle tissues. Bull Environ Contam Toxicol 38:1000-1005.

Bryan AM, Stone WB, Olafsson PG (1987b) Disposition of toxic PCB congeners in snapping turtle eggs: Expressed as toxic equivalents of TCDD. Bull Environ Contam Toxicol 39:791-796.

Carr A (1952) Handbook of turtles. Cornell Univ Press, Ithaca, NY.

Clark DB, Gibbons JW (1969) Dietary shift in the turtle *Trachemys scripta* (Schoepff) from youth to maturity. Copeia 1969(4):704-706.

Clark DR Jr, Krynitsky AJ (1980) Organochlorine residues in eggs of loggerhead and green sea turtles nesting at Merritt Island, Florida—July and August 1976. Pestic Monit J 14:7-10.

Clark DR Jr, Krynitsky AJ (1985) DDE residues and artificial incubation of loggerhead sea turtle eggs. Bull Environ Contam Toxicol 34:121-125.

Clark T, Clark K, Patterson S, Mackay D, Norstrom RJ (1988) Wildlife monitoring, modeling, and fugacity. Environ Sci Technol 22:120-127.

Collins, JT (1990) Standard, common, and current scientific names for North American amphibians and reptiles, 3rd ed. SSAR Herpetol Cir 19:1-41.

Conant R, Collins JT (1991) A field guide to reptiles and amphibians: Eastern and central North America. Houghton Mifflin, Boston, MA.

Cosgrove GE (1965) The radiosensitivity of snakes and box turtles. Radiat Res 25:706-712.

Cosgrove GE (1971) Reptilian radiobiology. J Am Vet Med Assoc 159:1678-1684.
Dalrymple GH (1977) Intra specific variation in the cranial feeding mechanism of turtles of the genus *Trionyx* (Reptilia, Testudines, Trionychidae). J Herpetol 11: 255-285.
Dobie JL, Ogren LH, Fitzpatrick JF Jr (1961) Food rates and records of the Atlantic ridley turtle (*Lepidochelys kempi*) from Louisiana. Copeia 1961:109-110.
Dunaway PB, Lewis LL, Story JD, Payne JA, Inglis JM (1969) Radiation effects in the Soricidae, Cricetidae and Muridae. In: Nelson DJ, Evans FC (eds), Proceedings 2nd national symposium on radioecology. Ann Arbor, MI, May 15-17, 1967, pp 173-184 (USAEC report CONF-670503).
Eisenberg JF, Frazier J (1983) A leatherback turtle (*Dermochelys coriacea*) feeding in the wild. J Herpetol 17:82-86.
Eisler R (1985a) Cadmium hazards to fish, wildlife, and invertebrates: A synoptic review. Biological rept 85(1.2), US Fish and Wildlife Service, Patuxent Wildlife Res Ctr, Laurel, MD.
Eisler R (1985b) Selenium hazards to fish, wildlife, and invertebrates: A synoptic review. Biological rept 85(1.5), US Fish and Wildlife Service, Patuxent Wildlife Res Ctr, Laurel, MD.
Eisler R (1986a) Chromium hazards to fish, wildlife, and invertebrates: A synoptic review. Biological rept 85(1.6), US Fish and Wildlife Service, Patuxent Wildlife Res Ctr, Laurel, MD.
Eisler R (1986b) Dioxin hazards to fish, wildlife, and invertebrates: A synoptic review. Biological rept 85(1.8), US Fish and Wildlife Service, Patuxent Wildlife Res Ctr, Laurel, MD.
Eisler R (1986c) Polychlorinated biphenyl hazards to fish, wildlife, and invertebrates: A synoptic review. Biological rept 85(1.7), US Fish and Wildlife Service, Patuxent Wildlife Res Ctr, Laurel, MD.
Eisler R (1987) Mercury hazards to fish, wildlife, and invertebrates: A synoptic review. Biological Rept 85(1.10), US Fish and Wildlife Service, Patuxent Wildlife Res Ctr, Laurel, MD.
Farris GC, Whicker FW, Dahl H (1969) Strontium-90 levels in mule deer and forage plants. In: Nelson DJ, Evans FC (eds), Proceedings 2nd national symposium on radioecology. Oak Ridge, TN, May 15-17, 1967, pp 602-608 (USAEC rept CONF-670503).
Fendley TT, Manlove MN, Brisbin, IL Jr (1977) The accumulation and elimination of radiocesium by naturally contaminated wood ducks. Hlth Phys 32:415-422.
Ferguson, DE (1963) Notes concerning the effect of heptachlor on certain poikilotherms. Copeia 1963:441-443.
Flickinger EL, King KA (1972) Some effects of aldrin-treated rice on Gulf Coast wildlife. J Wildl Mgt 36:706-727.
Flickinger EL, Mulhern BM (1980) Aldrin persists in yellow mud turtle. Herpetol Rev 11:29-30.
Folkerts GW (1968) Food habits of the stripe-necked musk turtle, *Sternotherus minor peltifer* Smith and Glass. J Herpetol 2:171-173.
Gibbons JW (1968) Growth rates of the common snapping turtle, *Chelydra serpentina*, in a polluted river. Herpetologica 24:266-267.
Gibbons JW (1987) Why do turtles live so long? Bioscience 37:262-269.
Gibbs W, Wilson E, Hodges H, Lushbaugh CC (1964) Comparative study of thyroid and whole body retention of iodine in box turtles. In: 2nd annual Oak

Ridge radioisotope conference. Oak Ridge Nat Lab, Oak Ridge, TN, pp 20–22 (USAEC rept TID-7689).

Goyer RA (1991) Toxic effects of metals. In: Amdur MO et al. (eds), Casarett and Doull's toxicology: The basic science of poisons, 4th ed. Pergamon Press, New York, pp 623–680.

Graham TE, Perkins RW (1976) Growth of the common snapping turtle, *Chelydra s. serpentina*, in a polluted marsh. Bull Md Herpetol Soc 12:123–125.

Halford DK, Markham OD, White GC (1983) Biological elimination rates of radioisotopes by mallards contaminated at a liquid radioactive waste disposal area. Hlth Phys 45:745–756.

Hall RJ (1980) Effects of environmental contaminants on reptiles: A review. US Fish and Wildlife Service special scientific rept—wildlife no 228, US Dept of the Interior, Washington, DC.

Hebert CE, Glooschenko V, Haffner GD, Lazar R (1993) Organic contaminants in snapping turtles (*Chelydra serpentina*) populations from southern Ontario, Canada. Arch Environ Contam Toxicol 24:35–43.

Helwig DD, Hora ME (1983) Polychlorinated biphenyl, mercury, and cadmium concentrations in Minnesota snapping turtles. Bull Environ Contam Toxicol 30: 186–190.

Hillestad HO, Reimold RJ, Stickney RR, Windom HL, Jenkins JH (1974) Pesticides, heavy metals and radionuclide uptake in loggerhead sea turtles from Georgia and South Carolina. Herpetol Rev 5:75.

Hinton TG, Scott DE (1990) Radioecological techniques used in herpetology with an emphasis on freshwater turtles. In: Gibbons JW (ed), Life history and ecology of the slider turtle. Smithsonian Institution Press, Washington, DC, pp 267–287.

Hoffman DJ, Rattner BA, Hall RJ (1990) Wildlife toxicology. Environ Sci Technol 24:276–283.

Holcomb CM, Jackson CG Jr, Jackson MM, Kleinbergs S (1971) Occurrence of radionuclides in the exoskeleton of turtles. In: Nelson DJ (ed), Proceedings 3rd national symposium on radioecology. Oak Ridge, TN, May 10–12, 1971, pp. 385–389 (USAEC rept CONF-710501-P1).

Holcomb CM, Parker WS (1979) Mirex residues in eggs and livers of two long-lived reptiles (*Chrysemys scripta* and *Terrapene carolina*) in Mississippi, 1970–1977. Bull Environ Contam Toxicol 23:369–371.

Jackson CG Jr, Holcomb CM, Kleinberg-Krisans S, Jackson MM (1974) Variation in strontium-90 exoskeleton burdens of turtles (Reptilia:Testudines) in southeastern United States. Herpetologica 30:406–409.

Jenkins JH, Monroe JR, Golley FB (1969) Comparison of fallout ^{137}Cs accumulation and excretion in certain southeastern mammals. In: Nelson DJ, Evans FC (eds), Proceedings 2nd national symposium on radioecology. Oak Ridge, TN, May 15–17, 1967, pp 623–626 (USAEC rept CONF-670503).

Kiviat E (1980) A Hudson River tidemarsh snapping turtle population. In: Transactions northeast fish wildlife conference. Ellenville, NY, April 27–30, 1980, pp 158–168.

Knight AW, Gibbons JW (1968) Food of the painted turtle, *Chrysemys picta*, in a polluted river. Am Midl Nat 80:559–562.

Kulp JL (1965) Radionuclides in man from nuclear tests. In: Fowler EB (ed), Radioactive fallout, soils, plants, food, man. Elsevier, Amsterdam, pp 247–284.

Lamb T, Bickham JW, Gibbons JW, Smolen MJ, McDowell S (1991) Genetic damage in a population of slider turtles (*Trachemys scripta*) inhabiting a radioactive reservoir. Arch Environ Contam Toxicol 20:138-142.

Mahmoud IY (1968) Feeding behavior in Kinosternid turtles. Herpetologica 24:300-305.

Mailhot H, Peters RH, Cornett RJ (1989) The biological half-time of radioactive Cs in poikilothermic and homeothermic animals. Hlth Phys 56:473-484.

Matsumura F (1975) Toxicology of insecticides. Plenum Press, New York.

McKim JM Jr, Johnson KL (1983) Polychlorinated biphenyls and p,p'-DDE in loggerhead and green postyearling Atlantic sea turtles. Bull Environ Contam Toxicol 31:53-60.

Meeks RL (1968) The accumulation of ^{36}Cl ring-labeled DDT in a fresh-water marsh. J Wildl Mgt 32:376-398.

Meyers-Schöne L (1989) Comparison of two freshwater turtle species as monitors of environmental contamination. PhD diss, Univ of Tennessee, Knoxville, TN.

Meyers-Schöne L, Shugart LR, Beauchamp JJ, Walton BT (1993) Comparison of two freshwater turtle species as monitors of chemical contamination: DNA damage and residue analysis. Environ Toxicol Chem 12:1487-1496.

Meylan A (1988) Spongivory in hawkbill turtles: A diet of glass. Science 239:393-395.

National Academy Press (1990) Decline of the sea turtles: Causes and prevention. Nat Academy Press, Washington, DC.

National Research Council (1986) Ecological knowledge and problem solving. Nat Academy Press, Washington, DC.

Olafsson PG, Bryan AM, Bush B, Stone W (1983) Snapping turtles—A biological screen for PCBs. Chemosphere 12:1525-1532.

Olie K, Schecter A, Constable J, Kooke RMM, Serne P, Slot PC, de Vries P (1989) Chlorinated dioxin and dibenzofuran levels in food and wildlife samples in the North and South of Vietnam. Chemosphere 19:493-496.

Ophel IL, Fraser CD (1971) The fate of cobalt-60 in a natural freshwater ecosystem. In: Nelson DJ (ed), Proceedings 3rd national symposium on radioecology. Oak Ridge, TN, May 10-12, 1971, pp 323-327 (USAEC rept CONF-710501-P1).

Overmann SR, Krajicek JJ (1989) The common snapping turtle: *Chelydra serpentina serpentina*, a bioindicator species for lead contamination in aquatic environments. Final rept submitted to Missouri Dept of Conservation.

Owen PJ, Wells MR (1976) Insecticide residues in two turtle species following treatments with DDT. Bull Environ Contam Toxicol 15:406-411.

Pearson JE, Tinsley K, Hernandez T (1973) Distribution of dieldrin in the turtle. Bull Environ Contam Toxicol 10:360-364.

Peters EL (1986) Radiocesium kinetics in the yellow-bellied turtle (*Pseudemys scripta*). MS thesis, Univ of Georgia, Athens, GA.

Peters EL, Brisbin IL Jr (1988) Radiocesium elimination in the yellow-bellied turtle (*Pseudemys scripta*). J Appl Ecol 25:461-471.

Phillips JB, Wells MR (1974) Adenosine triphosphatase activity in liver, intestinal mucosa, cloacal bladder, and kidney tissue of five turtle species following *in vitro* treatment with 1,1,1-trichloro-2,2-bis (p-chlorophenyl)ethane (DDT). J Agric Food Chem 22:404-407.

Punzo F, Laveglia J, Lohr D, Dahm PA (1979) Organochlorine insecticide residues in amphibians and reptiles from Iowa and lizards from the southeastern United States. Bull Environ Contam Toxicol 21:842-848.

Rappe C, Buser HR (1980) In: Kimbrough RD (ed), Halogenated biphenyls, terphenyls, naphthalenes, dibenzodioxins and related products. Elsevier, Amsterdam, pp 41–76.

Rappe C, Buser HR, Stalling DL, Smith LM, Dougherty RC (1981) Identification of polychlorinated dibenzofurans in environmental samples. Nature 292:524–526.

Reeves RG, Woodham DW, Ganyard MC, Bond CA (1977) Preliminary monitoring of agricultural pesticides in a cooperative Tobacco Pest Management Project in North Carolina, 1971 — first year study. Pestic Monit J 11:99–106.

Robinson KM, Wells MR (1975) Retention of a single oral dose of cadmium in tissues of the softshell turtle, *Trionyx spinifer*. Bull Environ Contam Toxicol 14:750–752.

Rosene W Jr, Stewart P, Adomaitis V (1961) Residues of heptachlor epoxide in wild animals. Proc Annu Conf Southeast Assoc Game Fish Comm 15:107–113.

Ryan JJ, Lou BPY, Hardy JA, Stone WB, O'Keefe P, Gierthy JF (1986) 2,3,7,8-tetrachlorodibenzo-p-dioxin and related dioxins and furans in snapping turtles (*Chelydra serpentina*) tissues from the upper St. Lawrence River. Chemosphere 15:537–548.

Scott DE, Whicker WW, Gibbons JW (1986) Effect of season on the retention of ^{137}Cs and ^{90}Sr by the yellow-bellied slider turtle (*Pseudemys scripta*). Can J Zool 64:2850–2853.

Shellabarger CJ, Gorbman A, Schatzlein FC, McGill D (1956) Some quantitative and qualitative aspects of I^{131} metabolism in turtles. Endocrinology 59:331–339.

Stalling DL, Smith LM, Petty JD, Hogan JW, Johnson LJ, Rappe C, Buser HR (1983) Residues of polychlorinated dibenzo-p-dioxins and dibenzofurans in Laurentian Great Lakes fish. In: Tucker RE, Young AL, Gray AP (eds), Human and environmental risks of chlorinated dioxins and related compounds. Plenum Press, New York, pp 221–240.

Stebbins RC (1985) A field guide to western reptiles and amphibians. Houghton Mifflin, Boston, MA.

Stickel LF (1951) Woodmouse and box turtle populations in an area treated annually with DDT for five years. J Wildl Mgt 15:161–164.

Stone WB, Kiviat E, Butkas SA (1980) Toxicants in snapping turtles. NY Fish Game J 27:39–50.

Stoneburner DL, Nicora MN, Blood ER (1980) Heavy metals in loggerhead sea turtle eggs (*Caretta caretta*): Evidence to support the hypothesis that demes exist in the western Atlantic population. J Herpetol 14:171–175.

Stribling HL, Brisbin IL Jr, Sweeney JR (1986) Radiocesium concentrations in two populations of feral hogs. Hlth Phys 50:852–854.

Suter GW, Loar JM (1992) Weighing the ecological risks of hazardous waste sites: The Oak Ridge case. Environ Sci Technol 26:432–438.

Talmage SS, Walton BT (1991) Small mammals as monitors of environmental contaminants. Rev Environ Contam Toxicol 119:47–145.

Thompson NP, Rankin PW, Johnston DW (1974) Polychlorinated biphenyls and p,p'-DDE in green turtle eggs from Ascension Island, South Atlantic Ocean. Bull Environ Contam Toxicol 11:399–406.

Towns AL (1987) ^{137}Cs and ^{90}Sr in turtles: A whole-body measurement technique and tissue distribution. MS thesis, Colorado State Univ, Fort Collins, CO.

Watson MR, Stone WB, Okoniewski JC, Smith LM (1985) Wildlife as monitors of the movement of polychlorinated biphenyls and other organic organochlorine compounds from a hazardous waste site. In: Transactions northeast fish wildlife conference. Hartford, CT, May 5-8, 1985, pp 91-104.

Wells MR, Phillips JB, Murphy GG (1974) ATPase activity in tissues of the map turtle, *Graptemys geographica*, following *in vitro* treatment with aldrin and dieldrin. Bull Environ Contam Toxicol 11:572-576.

Williams TA, Christiansen JL (1981) The niches of two sympatric softshell turtles, *Trionyx muticus* and *Trionyx spiniferus*, in Iowa. J Herpetol 15:303-308.

Yawetz A, Agosin M, Perry AS (1979) Metabolism of parathion and brain cholinesterase inhibition in four species of wild birds. Pest Biochem Physiol 11:294-300.

Yawetz A, Sidis I, Gasith A (1983) Metabolism of parathion and brain cholinesterase inhibition in aroclor 1254-treated and untreated Caspian terrapin (*Mauremys caspica rivulata*, Emididae, Chelonia) in comparison with two species of wild birds. Comp Biochem Physiol 75C:377-382.

Manuscript received September 1, 1993; Accepted September 20, 1993.

Index

Aldrin residues, turtles, 101
Alkylated PAHs as carcinogens, 34
Aluminum residues, turtles, 124
Aminoacenaphthene, carcinogen, 3
Aminoanthraquinone, carcinogen, 3
Aminochrysene, metabolic activation pathway, 14
Analytical chemistry, soil interactions, 72
Anthracene, carcinogen, 3
Aroclor® residues, turtles, 115
Aromatic amine structure/carcinogenicity relationship, 43
Atrazine, complexing fulvic acid carboxyl-binding sites, 75
Atrazine, complexing with fulvic acid, 76, 88
Atrazine, interactions with mineral soil, 81

Barium residues, turtles, 124
Bay-region dihydrodiol epoxide, illustrated, 11
Bay-region PAHs, metabolism to dihydrodiol-epoxides, 13
Benzo[a]pyrene, coal tar carcinogen, 2, 3
Benzo[a]pyrene, stereoselective metabolism, 12
Biomonitoring for carcinogenicity, PACs, 20

C-hydroxylation, influences arylamine carcinogenicity, 40
Cadmium residues, turtles, 124
Carbocation stabilities, PAHs, 33
Carcinogen, defined, 5
Carcinogenesis, chemical 3-stage process, 6

Carcinogenesis, multistage illustrated, 19
Carcinogenic activity of PACs, 3
Carcinogenic chemical substances, IARC Group 1, 2
Carcinogenic processes, IARC Group 1, 2
Carcinogenicity biomonitoring, PACs, 20
Carcinogenicity, polycyclic aromatic hydrocarbons (PACs), 1 ff.
Carcinogens involving PACs, 2
Chemical carcinogenesis of PACs, mechanisms, 5
Chemical carcinogenesis, 3-stage process, 6
Chemical carcinogenesis, initiation process, 6
Chemical carcinogenesis, principles, 5
Chemical carcinogenesis, progression, 6
Chemical carcinogenesis, tumor promotion, 6
Chlordane residues, turtles, 101
Chromium residues, turtles, 125
Chrysene, metabolic pathway, 9
Cigarette smoke, PACs carcinogenic effect, 2
^{60}Co, turtles, 135
Cobalt residues, turtles, 125
Copper residues, turtles, 125
Covalent binding index, DNA adducts and carcinogenicity, 20
^{137}Cs, turtles, 134
Cu^{2+} complexing on fulvic acid carboxyl-binding sites, 75
Cu^{2+}, soil aqueous slurry, 73
Cu^{2+}-fulvic acid chelation equilibria in soils, 80

DDT metabolites, turtles, 101
DDT residues, turtles, 99

Dibenz[*a,h*]anthracene, first pure carcinogen, 2
Dieldrin residues, turtles, 107
DNA adduct biomonitoring, carcinogenicity, 20
DNA adducts, detection method sensitivity table, 29
DNA adducts, determination techniques, 22, 28
DNA helix, *anti*-BPDE attached, illustrated, 36
DNA repair, role in chemical carcinogenesis, 7

Eggs, turtle, heavy metal residues, 124
Eggs, turtle, PCBs residues, 115
Eggs, turtle, pesticide residues, 101
Emydidae, freshwater turtles, 96
Endrin residues, turtles, 108
Environmental contaminants, turtles as monitors, 93 ff.
Epoxidation, influences arylamine carcinogenicity, 40
Epoxides, conversion to phenols/diols, 8
Epoxides/diol-epoxides structures, PAHs, 18
Epoxides, major aromatic double bond metabolism intermediates, 8

Fluoranthene, carcinogen, 3
Fluorescence spectroscopy, DNA adduct determination, 24, 26
Fulvic acid, complexing atrazine in soils, 74
Fulvic acid, weak acid polyelectrolyte, 72
Fulvic acid-Cu^{2+} chelation equilibria in soils, 80

Gas chromatography—mass spectrometry (GCMS), 25
GCMS, DNA adduct determination, 25, 26
GCMS, see Gas chromatography—mass spectrometry

Genotoxins, exposure measuring methods, 28
Glutathion, PAC metabolism, 8

Heavy metal residues, turtles, 123
Heptachlor epoxide residues, turtles, 109
Humic acid, solid phase in soils, 73

^{131}I, turtle thyroids, 141
Immunochemical methods, DNA adduct determination, 24, 28
Iron residues, turtles, 126

K-region epoxide formation, PACs, 9
K-region illustrated, 10
Kata-annelated PACs, 4
Kinetic rate constants, chemical speciation soil complexes, 78

L-region, PAHs, 10
Lead residues, turtles, 126
Lowest unoccupied π molecular orbital (LUMO), 36, 38
LUMO, see Lowest unoccupied π molecular orbital

Mercury residues, turtles, 129
Metal ions in soils, chemical speciation methods, 77
Metal ions, interactions with soils, 63 ff.
Methylchrysene, carcinogen, 3
Methylphenanthrene, carcinogen, 4
Microfiltration HPLC, chemical speciation soil complexes, 77
Mirex residues, turtles, 110
Mixed-function oxidase enzyme system, PAC metabolism, 8
Mixed-function oxidase systems, turtles, 143
Molybdenum residues, turtles, 131
Monitors of environmental contaminants, turtles, 93 ff.

Index

Multistage carcinogenesis, illustrated, 19
Mutagen, defined, 5

N-hydroxylation, influences arylamine carcinogenicity, 40
Nickel residues, turtles, 131
Nitro-PAHs, fragments responsible for mutagenicity, 43
Nonachlor residues, turtles, 111

Oncogenes, mechanistic role in multistage carcinogenesis, 7
Organic chemicals in soils, chemical speciation methods, 77
Organochlorine pesticide residues, turtles, 100

PAC metabolism, mixed-function oxidase system, 8
PACs, adduct formation, 13
PACs, adduct formation and carcinogenicity, 15
PACs, alkyl derivative carcinogens, 2
PACs, biophores mutagenic in *Salmonella*, 41
PACs, chemical structures, 5
PACs, conjugated π electron system, 4
PACs, contaminants in cancer-causing substances, 4
PACs, DNA adduct formation, 14, 15, 16
PACs, high-energy π-bonding orbital effects, 4
PACs, kata-annelated, 4
PACs, listed carcinogenic activity, 3
PACs, low-energy π^*-antibonding orbital effects, 4
PACs, mammalian detoxification system, 6
PACs, metabolic activation, 7
PACs, metabolic pathway, 9
PACs, mutagenic activity, 5
PACs, nitrogen containing, 12
PACs, pericondensed, 4
PACs, procarcinogens defined, 6
PACs, quantitative structure-activity relationships (QSARs), 27
PACs, see Polycyclic aromatic hydrocarbons
PACs, toxic effects molecular descriptors, 31
PAHs, alkylated as carcinogens, 34
PAHs, carbocation stabilities/solvation energies, 33
PAHs, carbonium ion calculations, 30
PAHs, diol-epoxides tumorigenic activity, 35
PAHs, epoxides/diol-epoxides structures, 18
PAHs, metabolism, 8
PAHs, metabolism, cytochrome P-450 system, 9
PAHs, methyl substitution effects on carcinogenicity, 11
PAHs, see Polycyclic aromatic hydrocarbons
Parathion poisoning, turtles, 143
PCBs residues, turtles, 114, 115
PCDDs residues, turtles, 119, 120
PCDFs residues, turtles, 119, 120
Pericondensed PACs, 4
Pesticide residues, turtles, 99
Pesticides, interactions with soils, 63 ff.
Phosphorothioate-^{35}S labeling, DNA adduct determination, 24, 28
Polychlorinated dibenzodioxins, see PCDDs
Polychlorinated dibenzofurans, see PCDFs
Polycyclic aromatic hydrocarbons (PACs), 1 ff.
Polycyclic aromatic hydrocarbons, carcinogenicity monitoring, 1 ff.
Polycyclic aromatic hydrocarbons, carcinogenicity prediction, 1 ff.
Polyelectrolyte, fulvic acid in soils, 72
Procarcinogens, PACs defined, 6
Protein adducts, determination techniques, 26, 28
Proto-oncogenes, tumor formation, 7
Pyrene, carcinogen, 4

Index

QSARs, predicting PACs carcinogenicity, 44
QSARs, see Quantitative structure-activity relationships
Quantitative structure-activity relationships (QSARs), 42

Radionuclides, turtles, 132
Risk assessment, DNA adduct monitoring carcinogenicity, 20
Risk assessment, protein adduct monitoring carcinogenicity, 25

Salmonella mutagenicity, molecules containing biophores, 41
Soil complexes, acid catalysis of hydrolysis in fulvic acid, 83
Soil complexes, chemical equilibria in mixed geochemical systems, 83
Soil complexes, intraparticle diffusion, 85
Soil complexes, weighted-average equilibrium functions, 82
Soil interactions, inner and outer variables, 67
Soil interactions, mathematical equilibria descriptions, 68
Soil interactions, mathematical kinetics description, 70
Soil interactions, stoichiometry in mixed geochemical systems, 72
Soil interactions with pesticides/metal ions, 63 ff.
Solvation energies, PAHs, 33
^{85}Sr, turtles, 139
^{90}Sr, turtles, 136
Stereoselective metabolism, benzo[*a*]pyrene, 12
Strontium residues, turtles, 131
Structure-activity relationships, PACs, 27

TCDD, see 2,3,7,8-Tetrachlorodibenzodioxin
2,3,7,8-TCDD residues, turtles, 122
2,3,7,8-Tetrachlorodibenzodioxin, 122
Toxaphene residues, turtles, 112
Toxic effects molecular descriptors, PACs, 31
Tumor suppression genes, neoplastic transformation, 7
Tumorigenic activity, PAHs diol-epoxides, 35
Turtle eggs, PCBs residues, 115
Turtle eggs, pesticide residues, 101
Turtles, as monitors, advantages/limitations, 144
Turtles, biochemical and histopathological stress responses, 142
Turtles, classification, 94 ff.
Turtles, common names, 95
Turtles, freshwater species, 95, 96
Turtles, growth rates vs contamination, 144
Turtles, heavy metal residues, 123
Turtles, marine species, 95, 98
Turtles, monitors of environmental contaminants, 93 ff.
Turtles, North American species, 95
Turtles, other biomonitoring tools, 142
Turtles, pesticide residues, 99
Turtles, radionuclide residues, 132
Turtles, scientific names, 95
Turtles, terrestrial species, 96, 99
Turtles, x-ray tolerance, 142

Ultrafiltration, chemical speciation of soil complexes, 77

X-rays, turtle tolerance, 142

Zinc residues, turtles, 131
^{65}Zn, turtles, 139

INFORMATION FOR AUTHORS

Reviews of Environmental Contamination and Toxicology

Edited by

George W. Ware

Published by

Springer-Verlag New York • Berlin • Heidelberg • Tokyo

The original copy and one good photocopy of the manuscript, complete with figures and tables, are required. Manuscripts will be published in the order in which they are received, reviewed, and accepted. They should be sent to the editor:

    Dr. George W. Ware
    Department of Entomology
    University of Arizona
    Tucson, AZ 85721
    Telephone and FAX: (602)299-3735

1. Manuscript

The manuscript, in English, should be typewritten, double-spaced throughout (including reference section), on one side of 8½ × 11-inch blank white paper, with at least one-inch margins. The first page of the manuscript should start with the title of the manuscript, name(s) of author(s), with author affiliation(s) as first-page starred footnotes, and "Contents" section. Pages should be numbered consecutively in arabic numerals, including those bearing figures and tables only. In titles, in-text outline headings and subheadings, figure legends, and table headings only the initial word, proper names, and universally capitalized words should be capitalized.

Footnotes should be inserted in text and numbered consecutively in the text using arabic numerals.

Tables should be typed on separate sheets and numbered consecutively within the text in *roman numerals*; they should bear a descriptive heading, in lower case, which is underscored with one line and which starts after the word "Table" and the appropriate roman numeral; *footnotes in tables* should be designated consecutively within a table by the lower-case alphabet. *Figures* (including photos, graphs, and line drawings) should be numbered consecutively within the text in arabic numerals; each figure should be affixed to a separate page bearing a legend (below the figure) in lower case starting with the term "Fig." and a number.

2. Summary

A concise but informative summary (double-spaced) must conclude the text of each manuscript; it should summarize the significant content and major conclusions presented. It must not be longer than two 8½ × 11-inch pages of double-spaced typing. As a summary, it should be more informative than the usual abstract.

3. References

All papers, books, and other works cited in the text must be included in a "References" section (*also double-spaced*) at the end of the manuscript. If comprehensive papers on the same subject have been published, they should be cited when the bibliographic citations extend farther back than to these papers.

All papers cited in the text should be given in parentheses and alphabetically when more than one reference is cited at a time, e.g. (Coats and Smith 1979, Holcombe et al. 1982, Stratton 1986), except when the author is mentioned, as for example, "and the study of Roberts and Stoydin (1985)." References to unpublished works should be kept to a minimum and mentioned only in the text itself in parentheses. References to published works are given at the end of the text in alphabetical order under the first author's name, citing all authors (surnames followed by initials throughout; do not use "and") according to the following examples:

Periodicals: Name(s), initials, year of publication in parentheses, full article title, journal title as abbreviated in "The ACS Style Guide: A Manual for Authors and Editors" of the American Chemical Society, volume number, colon, first and last page numbers. Example:

Leistra MT (1970) Distribution of 1,3-dichloropropene over the phases in soil. J Agric Food Chem 18:1124–1126.

Books: Name(s), initials, year of publication in parentheses, full title, edition, volume number, name of publisher, place of publication, first and last page numbers. Example:

Gosselin R, Hodge H, Smith R, Gleason M (1976) Clinical Toxicology of Commercial Products, 4th Ed. Wilkins-Williams, Baltimore, MD, pp 119–121.

Work in an edited collection: Name(s), initials, year of publication in parentheses, full title. In: name(s) and initial(s) of editor(s), the abbreviation ed(s) in parentheses, name of publisher, place of publication, first and last page numbers. Example:

Metcalf RL (1978) Fumigants. In: White-Stevens J (ed) Pesticides in the environment. Marcel Dekker, New York. pp 120–130.

Abbreviations

A	acre	sec	second(s)
bp	boiling point	μg	microgram(s)
cal	calorie	μL	microliter(s)
cm	centimeter(s)	μm	micrometer(s)
cu	cubic (as in "cu m")	mg	milligram(s)
d	day	mL	milliliter(s)
ft	foot (feet)	mm	millimeter(s)
gal	gallon(s)	m$\underline{\text{M}}$	millimolar
g	gram(s)	min	minute(s)
ha	hectare	$\underline{\text{M}}$	molar
hr	hour(s)	mon	month(s)
in.	inch(es)	ng	nanogram(s)
id	inside diameter	nm	nanometer(s) (millimicron)
kg	kilogram(s)	$\underline{\text{N}}$	normal
L	liter(s)	no.	number(s)
mp	melting point	od	outside diameter
m	meter(s)	oz	ounce(s)
ppb	parts per billion (*μ*g/kg)	sp gr	specific gravity
ppm	parts per million (mg/kg)	sq	square (as in "sq m")
ppt	parts per trillion (ng/kg)	vs	versus
pg	picogram	wk	week(s)
lb	pound(s)	wt	weight
psi	pounds per square inch	yr	year(s)
rpm	revolutions per minute		

Numbers: All numbers used with abbreviations and fractions or decimals are arabic numerals. Table numbers are roman numerals. Otherwise, numbers below ten are to be written out. Numerals should be used for a series (e.g., "0.5, 1, 5, 10, and 20 days"), for pH values, and for temperatures. When a sentence begins with a number, write it out.

Symbols: Special symbols (e.g., Greek letters) must be identified in the margin, e.g.

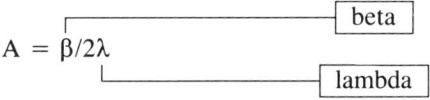

$$A = \beta/2\lambda$$

with labels: beta, lambda

Percent should be % in text, figures, and tables.

Style and format: The following examples illustrate the style and format to be followed (except for abandonment of periods with abbreviation):

Sklarew DS, Girvin DC (1986) Attenuation of polychlorinated biphenyls in soils. Reviews Environ Contam Toxicol 98:1–41.

Yang RHS (1986) The toxicology of methyl ethyl ketone. Residue Reviews 97:19–35.

References by the same author(s) are arranged chronologically. If more than one reference by the same author(s) published in the same year is cited, use a, b, c after year of publication in both text and reference list.

4. Illustrations

Illustrations may be included only when indispensable for the comprehension of text. They should not be used in place of concise explanations in text. Schematic line drawings must be drawn carefully. For other illustrations, clearly defined black-and-white glossy photos are required. Should darts (arrows) or letters be required on a photo or other type of illustration, they should be marked neatly with a soft pencil on a duplicate copy or on an overlay, with the end of each dart indicated by a fine pinprick; darts and lettering will be transferred to the illustrations by the publisher.

Photos should not be less than five × seven inches in size. Alterations of photos in page proof stage are not permitted. *Each photograph or other illustration should be marked on the back, distinctly but lightly, with soft pencil, with first author's name, figure number, manuscript page number, and the side which is the top.*

If illustrations from published books or periodicals are used, the exact source of each should be included in the figure legend: if these "borrowed" illustrations are copyrighted by others, permission of the copyright holder to reproduce the illustration must be secured by the author.

5. Chemical Nomenclature

All pesticides and other subject-matter chemicals should be identified according to *Chemical Abstracts*, with the full chemical name in text in parentheses or brackets the first time a common or trade name is used. *If many such names are used, a table of the names, their precise chemical designations, and their* Chemical Abstract Numbers (CAS) *should be included as the last table in the manuscript, with a numbered footnote reference to this fact on the first text page of the manuscript.*

6. Miscellaneous

Abbreviations: Common units of measurement and other commonly abbreviated terms and designations should be abbreviated as listed below; if any others are used often in a manuscript, they should be written out the first time used, followed by the normal and acceptable abbreviation in parentheses [e.g., Acceptable Daily Intake (ADI), Angstrom (Å), picogram (pg)]. Except for inch (in.) and number (no., when followed by a numeral), abbreviations are used without periods. Temperatures should be reported as "°C" or "°F" (e.g., mp 41° to 43°C). Because the metric system is the international standard, when pounds (lb) and gallons (gal) are used the metric equivalent should follow in parentheses.

7. Proofreading scheme

The senior author must return the Master set of page proofs to the Editor within one week of receipt. Author corrections should be clearly indicated on proof with ink, and in conformity with the standard "Proofreader's Marks" accompanying each set of proofs. In correcting proof, new or changed words or phrases should be carefully and legibly handprinted (not handwritten) in the margins.

8. Offprints

Senior authors receive 30 complimentary offprints of a published article. Additional offprints may be ordered from the publisher at the time the principal author receives the proof. Order forms for additional offprints will be sent to the senior author along with the page proofs.

9. Page charges

There are no page charges, regardless of length of manuscript. However, the cost of alteration (other than corrections of typesetting errors) attributable to authors' changes in the page proof, in excess of 10% of the original composition cost, will be charged to the authors.

If there are further questions, see any volume of *Reviews of Environmental Contamination and Toxicology* (formerly *Residue Reviews*) or telephone the Editor (see first page for telephone numbers). Volume 98 is especially helpful for style and format.